T0353591

A Beginner's Guide to Internet of Things Security

A Beginner's Guide to Internet of Things Security

Attacks, Applications, Authentication, and Fundamentals

B. B. Gupta
Aakanksha Tewari

CRC Press
Taylor & Francis Group
Boca Raton London New York

CRC Press is an imprint of the
Taylor & Francis Group, an **informa** business

CRC Press
Taylor & Francis Group
6000 Broken Sound Parkway NW, Suite 300
Boca Raton, FL 33487-2742

© 2020 by Taylor & Francis Group, LLC
CRC Press is an imprint of Taylor & Francis Group, an Informa business

No claim to original U.S. Government works

Printed on acid-free paper

International Standard Book Number-13: 978-0-367-43069-6 (Hardback)

This book contains information obtained from authentic and highly regarded sources. Reasonable efforts have been made to publish reliable data and information, but the author and publisher cannot assume responsibility for the validity of all materials or the consequences of their use. The authors and publishers have attempted to trace the copyright holders of all material reproduced in this publication and apologize to copyright holders if permission to publish in this form has not been obtained. If any copyright material has not been acknowledged please write and let us know so we may rectify in any future reprint.

Except as permitted under U.S. Copyright Law, no part of this book may be reprinted, reproduced, transmitted, or utilized in any form by any electronic, mechanical, or other means, now known or hereafter invented, including photocopying, microfilming, and recording, or in any information storage or retrieval system, without written permission from the publishers.

For permission to photocopy or use material electronically from this work, please access www.copyright.com (http://www.copyright.com/) or contact the Copyright Clearance Center, Inc. (CCC), 222 Rosewood Drive, Danvers, MA 01923, 978-750-8400. CCC is a not-for-profit organization that provides licenses and registration for a variety of users. For organizations that have been granted a photocopy license by the CCC, a separate system of payment has been arranged.

Trademark Notice: Product or corporate names may be trademarks or registered trademarks, and are used only for identification and explanation without intent to infringe.

Visit the Taylor & Francis Web site at
http://www.taylorandfrancis.com

and the CRC Press Web site at
http://www.crcpress.com

Dedicated to my parents and family for their constant support during the course of this book.

-B. B. Gupta

Dedicated to my mentor, my parents, and my friends for their constant encouragement and belief during the course of this book.

-Aakanksha Tewari

Contents

Preface

The potential capabilities of Internet of Things (IoT) can reduce a lot of time and expenditure of various organizations. These devices are excellent data collectors and sensors; therefore, they can help in efficient decision-making in a wide range of applications. However, security remains the biggest issue in the IoT domain. A lot of research is being carried out in this area to provide strong security and privacy mechanisms in IoT networks. The development of standards and protocol sets is necessary to build the IoT network properly. Only time will ultimately tell how far IoT will reach and how it will reshape the world. However, by the planned integration of existing technologies, we can make IoT networks secure and more efficient. We address various issues in securing IoT networks, which enabled us to develop various mutual authentication protocols that strengthen the security and privacy of IoT devices and prevent confidential data from theft. The present scenario of IoT research is mainly focused on the development of technologies for its implementation. By examining the recent statistics and literature, it also uncovers various challenges that have the potential to prevent IoT from growing to its full potential.

Specifically, the chapters contained in this book are summarized as follows:

Chapter 1: Evolution of Internet of Things (IoT): History, Forecasts, and Security. This chapter introduces the origin and concept of IoT along with its system requirements. It also explores the significance of security in the IoT domain by providing an in-depth statistical analysis of past events and future predictions.

Chapter 2: IoT Design, Standards, and Protocols. This chapter provides an overview of design and system requirements of IoT networks and security and privacy issues linked to the architecture. It also elaborates the configuration and underlying standards and protocols necessary for the development of IoT. This chapter also highlights the security requirements for platform development.

Chapter 3: IoT's Integration with Other Technologies. This chapter discusses various domains and the support they need from IoT for their growth. It also illuminates the security aspects of the integration of IoT with other domains such as cloud computing and big data.

Chapter 4: Industrial Internet of Things (IIoT). This chapter discusses the applications of IoT in industries. It explores various technologies such as edge computing, mobile technologies, and 3D printing, which have helped in realizing IIoT. We also discuss cybersecurity and the modular architecture of IIoT and how security comes into play with each module.

Chapter 5: Trust and Privacy in IoT. This chapter facilitates the significance of trust and privacy in IoT and how the lack of any one of these can diminish users' faith, which may result in the failure of any technology. It elaborates on the types of attacks and the degree of damage they can do to the networks. This chapter also gives an overview of methodologies through which trust can be ensured within a network.

Chapter 6: Authentication Mechanisms for IoT Networks. This chapter discusses how authentication mechanisms can secure IoT networks. It illuminates the workflow of authentication schemes customized for IoT and their feasibility and cost analysis.

Chapter 7: Provable Security Models and Existing Protocols. Further exploring the authentication-based security schemes, this chapter discusses some provable security models and existing protocols that have used these schemes, as they are crucial in ensuring that a protocol is secure.

Chapter 8: An Internet of Things (IoT)-Based Security Approach Ensuring Robust Location Privacy for the Healthcare Environment. This chapter discusses a simple low-cost authentication protocol, which ensures strong location privacy by giving proof of indistinguishability and forward secrecy. Further, it elaborates on the security strength of the protocol by a game-based model.

Acknowledgments

Writing a book is a huge task and more rewarding than one could fathom. This book entitled *A Beginner's Guide to Internet of Things Security* is the result of great contributions and encouragement from many people. None of this would have been possible without their ideas and support, which has helped greatly in enhancing the quality of this book. The authors would like to acknowledge the incredible CRC Press/Taylor & Francis Group staff, particularly Randi Cohen and her team, for their continuous assistance and motivation. This book would not have been possible without their technical support. The authors are eternally grateful to their families for their love and unconditional support at all times. In the end, the authors are most thankful towards the Almighty who is always helping us to overcome every obstacle not only for this work but also throughout our lives.

September 2019
B. B. Gupta
Aakanksha Tewari

Authors

B. B. Gupta received a Ph.D. degree from the Indian Institute of Technology Roorkee, India, in the area of information and cybersecurity. He has published more than 200 research papers in international journals and conferences of high repute, including IEEE, Elsevier, ACM, Springer, Wiley, Taylor & Francis, Inderscience, etc. He has visited several countries, that is, Canada, Japan, Malaysia, Australia, China, Hong-Kong, Italy, and Spain, to present his research work. His biography was selected and published in the 30th edition of Marquis Who's Who in the World, 2012. Dr. Gupta also received the Young Faculty Research Fellowship Award from the Ministry of Electronics and Information Technology, Government of India, in 2018. He is also working as a principal investigator of various R&D projects. He is serving as an associate editor of IEEE Access and IEEE TII, and an executive editor of IJITCA and Inderscience, respectively. At present, Dr. Gupta is working as an assistant professor in the Department of Computer Engineering, National Institute of Technology, Kurukshetra, India. His research interests include information security, cybersecurity, mobile security, cloud computing, web security, intrusion detection, and phishing.

Aakanksha Tewari is a Ph.D. scholar in the Department of Computer Engineering at the National Institute of Technology (NIT), Kurukshetra, India. Her research interests include computer networks, information security, cloud computing, phishing detection, Internet of Things, radio frequency identification (RFID) authentication, and number theory and cryptography. She has done her M. Tech. (Computer Engineering) from the Department of Computer Engineering at the National Institute of Technology (NIT), Kurukshetra, India. She has participated and won in various national workshops and poster presentations. Currently, her research work is based on security and privacy in IoT networks and mutual authentication of RFID tags.

Evolution of Internet of Things (IoT)

History, Forecasts, and Security

1

The Internet of Things (IoT) is a new paradigm which is transforming everything from the consumer market, that is, household devices to industrial applications at large scales. The Internet was always intended to bring pieces of software, services, and people together on one platform at a global level [1]. Nowadays with the evolution of IoT, day-to-day objects have also become a part of the Internet sending and receiving updates continuously from one place to another. Therefore, we can define IoT as a network of interconnected devices, which provide services and share data-connecting and performing tasks in various applications [2].

The highly distributed and dynamic nature of IoT enables it to receive and store data continuously in huge amounts. For example, in the field of healthcare, it has led to remote health monitoring, emergency notifications, etc. The consumer electronics markets are also exploding with wearable gadgets [3]. Various domains such as wireless sensor networks (WSN), embedded systems, and radio frequency identification (RFID) are found to be huge contributors towards the growth of IoT.

As IoT is an evolving domain, it requires a lot of attention from the researchers and the industry as well. Various standardization organizations such as IEEE and IETF are also working towards developing standards and protocols for IoT architecture. The sensors and actuators that are consumed in the consumer electronics market are very low cost and small sized and have high computational capabilities, which are the reasons for the growth of IoT as automation is made so easy. Industries are also deploying IoT at large scales such as in retail management and transportation [3,4].

The understructure for IoT is the Internet providing connectivity, which also adds to the vulnerabilities in these networks. IoT networks face the same security threats as the Internet; in addition, due to their limited capabilities and simpler architecture, they are easier to compromise. At the physical layer, most of the IoT devices use RFID, therefore ensuring that RFID tags can secure our data from any threat to security and privacy [5].

Our aim is to perform an in-depth analysis of the recent advancements in the field of security and privacy in IoT networks. Research needs to be done in order to facilitate the integration of IoT with other technologies in a secure environment. This can be accomplished by designing standard communication methodologies and standard protocols. It is a primary requirement to make IoT power efficient and reliable. The use of proper authentication mechanisms is one way to ensure security against various attacks and maintain the availability and integrity of data and services at all times for authorized users.

1.1 EVOLUTION OF IoT

In order to grasp the concept of IoT, it is important that we dig a little deeper towards the history of IoT and how it became what it is today. During 1982, a group of programmers at Carnegie Mellon University were successful in connecting a soda-dispensing machine to the Internet, which was able to check whether the machine has any cold soda cans left before going to purchase. In 1990, John Romkey connected a toaster to the Internet through which he was able to turn it on and off. These are the few earliest examples of automation before IoT. The year 1999 brought a big reform to consumer electronics when the term "IoT" was coined by Kevin Ashton, director of Massachusetts Institute of Technology s auto-ID center [6,7].

The first Internet-connected refrigerator was introduced in 2000. In 2004, the term "IoT" was used by magazines such as The Guardian. During the same period, the RFID technology was also being deployed on a large scale in various industries such as Walmart. During 2005, UN's (United Nations) international telecom union (ITU) stated that IoT is a new dimension of Information and Communication Technology (ICT), which will multiply the connections leading to a new dynamic network [8]. In 2007, the first iPhone was out in the market, which changed the whole scenario of communication between users. In 2010, the Chinese government declared IoT to be a key technology and a major part of China's long-term development [9].

In 2011, Nest developed a Wi-Fi-enabled thermostat, which was capable of modifying the atmosphere by monitoring users' habits. Platforms like Arduino were made more advanced so that they simplify IoT-based simulation for experimentation purposes [10]. In 2014, Amazon released Echo into the smart home domain. The research firm Gartner added IoT to their hype cycle, which is their graph that compares popular technologies. In 2016, Apple showcased HomeKit, which is a proprietary platform for supporting smart home appliance software. During the same year, the first IoT-related malware called Mirai was discovered, which collected default credentials through which it attacked devices using them. In 2017, governments started taking IoT security seriously as a result of several violations [11]. Various countries proposed a ban for hard-coded passwords.

In the past few years, IoT devices and related technologies have gotten a little cheaper and broadly accepted globally. In the long term, IoT is going to be the new normal. However, Gartner reports claim that IoT is going to hit a plateau in the coming years. Figure 1.1 shows a brief timeline for IoT development.

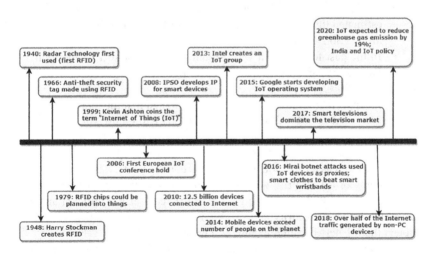

FIGURE 1.1 Brief history of IoT and RFID technology.

1.2 STATISTICS AND FORECASTS

To secure IoT devices, we need to incorporate security between the network connections and the software applications that run on those devices. It is anticipated that in the next ten years, a shift will happen from consumer-based IoT to industrial IoT. However, it is not sure whether the users and industry are fully aware of the impact that IoT is going to have in the next few years. According to Gartner's report, by 2020, 95% of consumer electronics will be enabled by IoT. As IoT networks are evolving into huge ecosystems, the government needs to devise some cybersecurity plans that can recognize threats and prevent them. Some of the recent trends that show the significance of IoT are as follows [10]:

- The number of Internet-connected devices surpassed the number of people on the planet in 2008, and it is estimated that by 2020, a number of Internet-connected devices will be around 50 billion.
- IoT is having a huge economic impact on the consumer electronics market; it is estimated that IoT technology will make about 19 trillion dollars of profit in the coming decade.
- It is estimated that by the end of 2019, about 2 billion home appliances will be shipped and will make homes a part of the Internet as well.
- Healthcare domain has adopted IoT at a large scale, and it is estimated that by 2025, the total worth of IoT in the markets will be around $6.2 trillion.
- In the past few years, it was estimated that about 10% of the total number of cars are connected to the Internet, and by 2020, it will reach 90%.
- A report by McKinsey & Company estimated that the total revenue generated by IoT will range from $4 to $11 trillion by 2025.

The aforementioned trends show that the rapid growth IoT has been in the past few years as well as its potential growth in the coming years. It is estimated that the economy of IoT security will be around $28.90 billion in 2020. However, in 2015, it was $6.89 billion. The growth in IoT requires a significant amount of investment in its security as well. We need security mechanisms that can protect the IoT network architecture as a whole [12,13].

The current rate of development in IoT technology will help us predict its future. Currently, the number of connected IoT devices is around 5 billion, most of which are personal devices. Most of the devices are consumer based, and the largest increase in number is shown by automotive. IoT has provided various business opportunities, such as smart homes and automobiles with better decision-making skills. However, we also need to address and resolve any security issues, which might prove to be a big threat to IoT networks in the future. Hewlett-Packard (HP) reported that about 70% of connected IoT devices these days are vulnerable to various threats due to their default settings. In a report by *Harvard Business Review*, it was estimated that nearly 45% of IoT networks have to deal with data privacy-related issues [14]. Figure 1.2 shows the rate of growth of IoT applications in the next ten years.

Every IoT device needs a unique identification technique so that there is no ambiguity in communication. The most popular identification technique these days is RFID, which has several advantages over the barcode. The RFID system consists of RFID tags, a reader, and a database or distributed databases, which keep records of the objects or devices within the network. RFID technology uses radio frequency mechanism for unique identification of objects. Every RFID tag contains a chip that has some storage, space, and sensing capability. In the last couple of years, significant growth in the use of RFID technology with various applications has given rise to various security threats such as a denial of service, man-in-the-middle, and desynchronization attacks.

To promote IoT and make larger networks, we must consider their security issues that affect the development of IoT networks and devices. Nowadays, attackers are always one step ahead of existing security mechanisms; they can

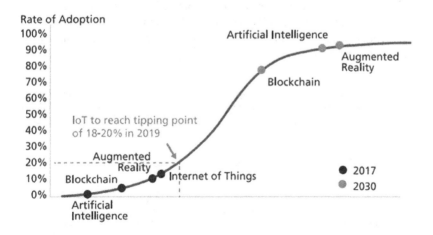

FIGURE 1.2 Percentage growth of IoT applications in the next ten years. (*Source*: DBS Bank.)

carry out attacks, which can disrupt services or transfer control to attackers at remote locations. IoT devices are vulnerable to various attacks such as replay, forgery, phishing, and denial of service.

In January 2015, Proofpoint revealed a spamming incident where the traffic was routed through several devices across various countries. This global attack had more than 750,000 malicious emails transferred from various locations, which were sent from consumer devices such as home routers, televisions, media and centers. Later on, it was discovered that at least one refrigerator was also involved in this attack. It was observed that the incident started from December 23, 2014, and continued till January 6, 2014, where the malicious email traffic was sent thrice a day with a burst size of 100,000 emails each. The targets were both enterprises and individuals. The primary cause of these attacks was a lack of caution and awareness. The attackers exploited misconfigurations and the continued use of default passwords, which made the devices vulnerable and easy to control [13–15].

Another wave of IoT attacks occurred in 2016, which mainly involved devices such as IP cameras and routers. The compromised devices were turned into botnets. These botnets were used collectively to launch attacks on a large scale. The cybercriminals are becoming more and more advanced. In an attack in 2018, a device that controlled around 15 CCTV cameras was attacked. However, in due time, the security operator detected the malicious activity and issued a warning that this might infect many more CCTV models. Another cause of these flaws is a lack of complete patching of IoT devices [16].

The IoT-based companies sometimes ignore security, or they are not experienced enough to realize the gravity of the situation (Figures 1.3 and 1.4). Lack of consumer awareness is also a very big cause behind these successful attacks. Consumers are often excited about the features and functions these devices provide so that they do not pay attention to security updates and setting strong passwords.

	2016	2017	2018	2030
IoT units installed base – total (m)	6,382	8,381	11,197	125,000
Consumer devices (m)	3,963	5,244	7,036	75,000
Consumer devices as a % of total devices	62%	63%	63%	60%
Connected devices per person	5	5	5	5
World population (m)	7,400	7,600	7,700	8,500
IoT adoption rate	11%	14%	18%	176%

FIGURE 1.3   Growth in IoT consumer devices. (*Source*: Gartner, Inc.)

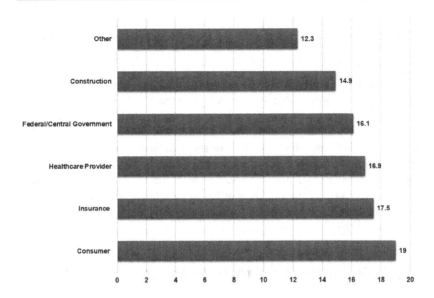

FIGURE 1.4 Top five IoT-based industries from 2017 to 2022.

The attacks are proof of the lack of security schemes in IoT networks, which need to be taken very seriously. In the current scenario, IoT gadgets are vulnerable to various attacks that may disrupt their services and transfer the control to some remote attacker. The attacker can impersonate a server and make the devices decrease their message-sending rates or increase the rate of their resource consumption and bandwidth. The attacker might also impersonate any tag and send multiple fake requests to engage servers' resources eventually leading to DoS.

IoT devices are also needed to be protected from a wide range of threats, which include malware infections, disruption of services, and information theft. The attacker could easily gain in controlling the devices that are a part of smart home, automobiles, or personal fitness and disease-monitoring gadgets. An attacker can simply hack the software in a person's smart watch or an insulin pump to track their location, or they might gain access to the information systems present in the automobiles and use them to carry malicious activities.

The most serious threat IoT devices face is malware such as Trojans, viruses, and worms that can disable IoT systems. Besides, this work also needs to be done to ensure that updates received by IoT devices are secure along with secure default settings. There is still a huge room for improvement when it comes to securing the IoT architecture.

1.3 FUNDAMENTALS OF IoT SECURITY

To implement IoT in real time, we need to integrate it with other existing technologies. However, due to the lack of standards, IoT does not have a fixed architecture yet. Various groups are working on developing protocols and standard modular or layered IoT architecture. The existing idea of IoT architecture has three layers (which in some cases are further subdivided): perception layer, network layer, and application layer. Each layer has its own security issues that need to be resolved to facilitate its growth. We need security mechanisms at every layer of IoT to prevent any security and privacy threats. In this section, we briefly discuss the security and system requirements of IoT networks.

1.3.1 Security at Different Layers

Each layer of IoT architecture has some security- and privacy-related issues, which we are required to be addressed in order to secure IoT applications. In fact, all such issues should be taken into account and remedied at the very initial stage of system design. The existing IoT architecture raises the requirement of proper security checks during the beginning and at regular intervals for an IoT network as a whole. At the network layer, threats to confidentiality, integrity, and availability should be dealt with. Attacks such as eavesdropping, man-in-the-middle, DoS/DDoS, and network intrusion are common threats at this layer. The application layer needs different security standards as per the application requirements, which makes the task of securing the application hard and complicated.

1.3.2 System Requirements of the IoT System

In order to ensure the security of an IoT system, we can ensure the basic security requirements at the RFID level. These are considered to be basic criteria that must be fulfilled by the protocol to protect security attacks on a system [17].

- *Mutual authentication*: It is an essential requirement that the server should authenticate the tag to be legitimate and tag should also authenticate the server before they exchange any important information.

- *Anonymity*: It refers to concealing the identity from any adversary by an encryption technique. It is said to be strong if the attacker is not able to trace back the tag with the help of its communication with the tag.
- *Forward secrecy*: It ensures that some compromised present communication information could not be used to obtain past communication information.
- *Confidentiality*: It ensures that all the information is transferred secretly without getting revealed to anyone except the receiver.
- *Availability*: It ensures that the authentication protocol works all the time, thus not causing desynchronization between the tag and the server.

1.4 CONCLUSION

Our research work has addressed various issues in securing IoT networks, which enabled us to develop various mutual authentication protocols that strengthen the security and privacy of IoT devices and prevent confidential data from theft. We have discussed that the present scenario of IoT research is mainly focused on the development of technologies for its implementation. By examining the recent statistics and literature, we have also uncovered various challenges that have the potential to prevent IoT from growing to its full potential.

The potential capabilities of IoT can reduce a lot of time and expenditure of various organizations. These devices are excellent data collectors and sensors; therefore, they can help in efficient decision-making in a wide range of applications. However, security remains the biggest issue in the IoT domain. A lot of research is being carried out in this area to provide strong security and privacy mechanisms in IoT networks. The development of standards and protocol set is necessary to build the IoT network properly. Only time will ultimately tell how far IoT will reach and how it will reshape the world. However, by the planned integration of existing technologies, we can make IoT networks secure and more efficient.

IoT Design, Standards, and Protocols

2

The Internet of Things (IoT) is a highly distributed and dynamic cyber-physical system. It integrates devices having sensors, identification systems, storage, communication, processing, and networking capabilities. With the advancement of technology, sensors and actuators have more complex specifications, which are available in lesser cost and smaller sizes nowadays. These devices are making the growth of ubiquitous computing easier. Various industries are deploying IoT for the development of industrial applications to increase automation and monitoring. The rapid advancement of technology and industrialization will enable the applications of IoT in various fields and industries. For example, consider the food industry that has integrated radio frequency identification (RFID) technology with wireless sensor networks (WSN) in order to automate the process of monitoring, tracking, and measuring the quality of food of any food supply chain [1,2].

In this chapter, we will survey the current scenario of security and trust management in IoT by analyzing existing works and taxonomies of security schemes and checking their compatibility with the existing IoT applications. We also open issues and challenges and the expected future trends related to IoT growth and need of security. The ongoing research in the field of IoT is majorly focused on technology [3]. The full realization of IoT is not done yet; therefore, there are huge opportunities of technical growth and development in the field of IoT. However, the rapid growth rate of technology and the research in IoT will have applications in the fields such as law, economics, management, and social studies [4]. RFID technology, which is a primary enabler for IoT, has also seen a rapid growth in the last couple of years. It has applications in the field of retail management, transportation and logistics, and healthcare [8]. With the increasing use of RFID, the maintenance of security and data privacy has also become a chief concern.

RFID systems are always deployed in bulk; these systems comprise a set of tags that have some storage and computational capabilities. These tags are monitored and identified using readers equipped with transponders to communicate with these tags. RFID systems are being used in retail at large scales replacing barcodes due to their resistance to tampering and ease of reading. RFID tags do not have to be in the line of sight of the reader for identification and communication. However, RFID tags have a limited storage and computational capability [9,10].

2.1 LAYERED IoT ARCHITECTURE

The standardization organizations are working on designing layered architecture for IoT. However, as of now, IoT devices have no standard architecture; there are few tentative models that are being used by the researchers. The existing IoT architectures have three, four, or five layers; the most widely used is the three-layered architecture: application layer, middleware layer, and sensor layer, as shown in Figure 2.1.

Perception/sensor layer: This layer identifies objects uniquely within the IoT networks. It also uses sensors to collect data from its surroundings and other objects. This layer deploys WSN or RFID tags depending upon the applications' requirements. The data collected by the sensors is transferred to the middleware layer for further processing.

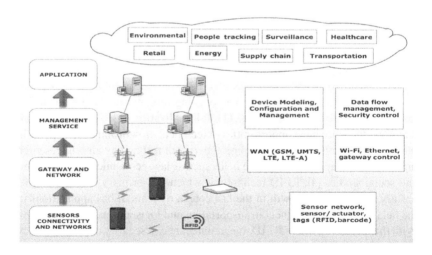

FIGURE 2.1   IoT architecture.

Middleware layer: This layer provides a set of communication protocols and network support methodologies to IoT devices. In the existing literature, this layer has also been divided into two or more layers each providing different services depending upon the requirements of IoT network. In some four-layered architectures, we have a processing layer that collects and processes data from the sensor layer; the processed data is then sent to a transport layer, which provides access to communication technologies such as Bluetooth and Wi-Fi to send and receive data, as shown in Figure 2.1. This layer also assigns unique addresses to objects within a network using IPv6 addressing [12–15].

Application layer: This layer receives processed data from its subsequent layers and uses it according to the applications' requirements. Some architectures divide this layer into two parts: the business layer deals with the security and privacy requirements of an application and manages the data, and the upper application layer deals with interoperability and distinguishes among various applications.

2.2 SECURITY AND PRIVACY ISSUES WITH IoT ARCHITECTURE

In order to ensure security in IoT applications, our primary aim is to address the security- and privacy-related issues at each layer of IoT architecture. The existing architectures require strong security checks and traffic monitoring at regular intervals. In this section, we discuss the security issues that exist at each layer of the IoT architecture.

2.2.1 Perception-Layer Security Problems

The primary technologies used in the perception layers are WSN, RFID, and other types of sensing and identification techniques. The most common types of threats faced by this layer are as follows:

a. *Node capture*: The nodes present at the network gateway are more likely to be compromised, which may result in the leakage of important information that poses a serious threat to the security of the entire network.

b. *Fake node and malicious data*: The adversary can add a malicious node to the existing system through which they can circulate malicious codes and information over the network, and can infect the whole system.

c. *Replay attack*: The adversary replays a previous message to the destination node so as to compromise the network trust and authentication schemes.

2.2.2 Network-Layer Security Problems

At the network layer, we have to deal with threats to confidentiality, integrity, and availability. Attacks such as eavesdropping, man-in-the middle, DoS/DDoS, and network intrusion are common threats at this layer.

a. *Heterogeneity*: Due to the use of different technologies and protocols, security and network coordination are hard to maintain, thus making the system vulnerable.

b. *Scalability issues*: IoT comprises a large number of devices within a network, and more devices may enter or leave the network at different times, which raises issues such as a lack of authentication and network congestion. It also depletes a lot of resources.

c. *Data disclosure*: Using social engineering techniques, the adversary might be able to obtain sensitive information from the network. The IoT devices collectively have huge amounts of data; therefore, using certain data retrieval techniques makes it easy to extract information from the nodes.

2.2.3 Application-Layer Security Problems

This layer needs different security standards as per the application's requirements, which makes the task of securing the application hard and complicated. Some of the security and privacy issues at this layer are as follows:

a. *Mutual authentication and node identification*: Each application has a different set of users, which requires various degrees of access privileges. Thus, to prevent any illegal access, effective authentication schemes should be applied.

b. *Information privacy*: User privacy should be ensured for each communication session. However, in some cases, the techniques that are being used to process data might be vulnerable, which leads to data loss. Over a long term, it can create huge damages to the system.

c. *Data management*: Due to huge data collections, the system complexity increases, which requires a lot of resources and complex algorithms for data management. It may also result in data loss.

d. *Application-specific vulnerabilities*: While developing modules for an application, some vulnerabilities might be left behind, which are unknown to the user. These can be exploited by the adversary later on.

2.3 IoT PROTOCOL DESIGN

IETF (Internet Engineering Task Force) has formed various groups with the task of developing protocols for IoT networks considering the storage and computation constraints of the devices. Another important thing to keep in mind is the ability to communicate with other devices having different architectures and applications [18].

2.3.1 Protocol Stack for IoT

IoT devices have a standard protocol stack, which can be divided into the following:

a. IEEE 802.15.4 underlines certain protocols that require low energy for communication at the physical layer, thus laying down a set of rules for communication at the physical layer and a base for higher layers [19].

b. The 6LoWPAN layer enables IPv6 packet transmission, thus using 102 bytes for low-energy protocols at higher layers. This layer also incorporates mechanisms for packet fragmentation and reassembly [20].

c. Routing Protocol for Low Power and Lossy Networks (RPL) support routing in 6LoWPAN. It also enables an adaptable architecture to support the routing and optimization requirements for IoT applications [20,21].

d. IETF work groups are currently developing Constrained Application Protocol (CoAP) for application-layer communication. It will also ensure interoperability in case of communication between heterogeneous networks or devices [22].

2.3.2 Security Requirements

The security of IoT networks should be addressed by the communication protocols used at each layer rather than the application of external security models. The primary security designs aim to ensure confidentiality, integrity, authentication, and anonymity. The security is ensured with the help of modes, access control functions, and time synchronization present in IEEE 802.15.4.

For the 6LoWPAN layer, there are no security mechanisms defined so far. However, the related RFCs mention security challenges and usage of security in the network layer. For example, RFC 6606 discusses the significances of security designs with time synchronization and localization. The RPL protocol incorporates security modes. The control message contains 4-byte security field. The field code has higher-order bit value to specify whether or not security is implemented for the given message [20–23].

CoAP supports security at the application layer in integration with DTLS (Datagram Transport Layer Security). DTLS ensures confidentiality, integrity, and authentication at the application level. Figure 2.2 shows the security designs at protocol level for each layer.

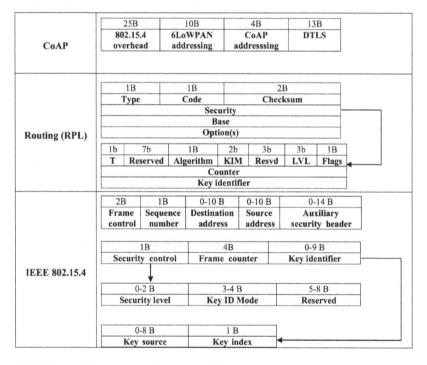

FIGURE 2.2   Security data formats at different layers.

2.4 IETF AND IEEE DESIGN STANDARDS

In this section, we describe the standards that aid the realization of IoT and maintenance of security and privacy of these networks. Nowadays, protocols and architectures are being developed in order to extend Internet technologies for low-cost devices. These protocols are developed by modifying the existing communication protocols. Several protocols have been developed at the network layer to ensure proper communication and data flow. Protocols such as IPv6 over low-power WPANs have been developed for the transmission of packets and so on. These new standards provide a support for the growth of IoT networks, making end-to-end connectivity of constrained devices and day-to-day objects possible.

It may seem like it but global connectivity is not the primary reason behind the success of the Internet, the web services, and the World Wide Web. For the IoT networks, we need web services to be implemented on embedded systems to exploit all the applications that IoT has to offer in order to build a Web of Things. Recently, the working group IETF Constrained RESTful Environments (CoRE) has begun working on developing standards to enable the integration of IoT devices with Internet and provide a related set of services. For application development, high-level programming standards are required for the implementation of Semantic Web of Things or building a sensor network. With the new technologies and standards, deploying a sensor network that can organize itself has become possible. These networks can be connected with each other using IPv6 Internet. In order to develop applications, we need standard interfaces to enable data processing and communication and access sensor-collected data. The IETF working group have developed various novel standards and protocols.

The key realizations of the related IETF working groups will be discussed, complemented with an extensive overview of related research results that illustrate how these novel technologies can be extended or used to tackle other problems, and complemented with a discussion on open issues and challenges.

2.5 TAXONOMY OF THREATS TO IoT NETWORKS

Considering the attack patterns, we present an attack mechanism (shown in Figure 2.3) followed by the adversary while planning an IoT attack (or any other attack campaign). There are four steps taken by the attacker while

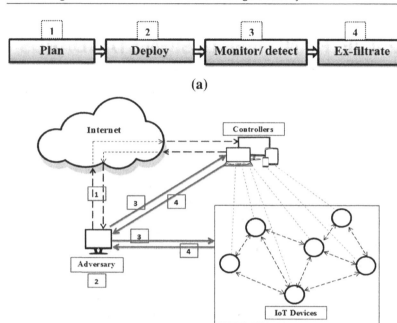

FIGURE 2.3   (a) Stages in an attack. (b) IoT attack mechanism.

carrying out IoT attack campaign (which can be generalized for any security attack) that is mentioned below:

- *Plan*: The first step taken by the adversary is planning of the attack, which is not specific to IoT only. Before carrying out an attack, the first step is to gain as much information as possible about the network that is to be exploited. The users can gain information by hearing the messages between the IoT devices and the server or between the devices.
- *Deploy*: After collecting sufficient information, the attacker plans the attack campaign, that is, deciding the medium of attack, getting access to the IoT devices to install malicious software, and extracting information using a remote system.
- *Monitor/detect*: After the malware is successfully installed in the devices of an IoT network, the next step is to monitor the communication and other functionalities of the IoT objects and gather the required confidential information or bring the network down, etc.

TABLE 2.1   Attacks on IoT Networks

ATTACK	DESCRIPTION
Physical attack	The attacker may compromise several IoT devices, which have a fixed topology; this attack is at the physical layer, and the breaches cause data to be stolen.
Unauthorized access	IoT devices may have some vulnerability or improper configuration, and the attacker exploits them to gain illegal access to the devices or the network.
Cloning attack	As the IoT nodes have a fairly simple architecture, it is possible to make their duplicates, and it harms the network integrity.
Brute force	Due to limited resources and computational power, it is easy to perform a brute-force attack on IoT devices to compromise them and retrieve information.
Denial of Service (DoS) attack	Due to restricted resources, DoS attacks can be easily performed on IoT networks to refrain the nodes from the services and resources.
Desynchronization attack	These attacks disturb the synchronization between the RFID tag and the server; these are performed by continuously intercepting the communication and message exchanges between two parties.
Full disclosure attack	This attack enables the attacker to retrieve the identity and other confidential information of an RFID tag.

- *Exfiltrate*: After fulfilling their objectives, the adversary's next aim is to exfiltrate out of the network without being detected to end their campaign successfully.

The IoT devices are basically wireless devices with a limited number of resources and require the identification of other objects and humans as well. They are widely accepted in every field these days. But, having a limited functionality, we require some lightweight mechanisms to secure them from various attacks, as given in Table 2.1.

2.6 CONCLUSION

The popularity of sensor networks (and in a broader sense IoT) has increased significantly over the past ten years. Integration of these embedded devices into the Internet is challenging, since they have characteristics that differ strongly from traditional Internet devices, such as very limited energy, memory, and processing capabilities. Initially, research focused on developing proprietary solutions that were typically vendor-specific and did not allow end-to-end connectivity between client devices and sensor devices. However, the use of standardized protocols enables the integration of constrained devices in the IPv6 Internet, both at the network level and at the service level. We present a high-level overview of the ongoing IETF standardization work that focuses on enabling direct connectivity between clients and sensor devices. To this end, different IETF groups are currently active. The IETF 6LoWPAN and ROLL groups focus on the addressability and routing of IPv6, whereas the IETF CoRE group focuses on realizing an embedded counterpart for RESTful web services. By combining these protocols, an embedded protocol stack can be created that has similar characteristics to traditional Internet protocol stacks. In fact, the IETF protocols are designed to enable easy translation from Internet protocols to sensor protocols, and vice versa. We also discuss the combination of IETF protocols enabling a flexible, direct interaction between internet clients and embedded IoT devices. However, we show that the advent of standardized protocols is not an end point, but only a starting point for exploring additional open issues that should be solved to realize an all-encompassing IoT. Several open challenges remain, such as resource representation, security, dealing with sleeping nodes, energy efficiency, integration with existing web service technologies and tools, linking with cloud services, use of semantics, easy creation of applications, scalability, interoperability with other wireless standards, and maintainability. Anyone involved in IoT research (whether dealing with network-layer aspects or service-layer aspects) will, sooner or later, be confronted with the IETF protocols.

IoT's Integration with Other Technologies

3

In the early years of development, Internet of Things (IoT) had fewer resources and machines, which had high-design installation cost. However, in the past few years, there had been a significant growth in this domain that is making the resources cheaper, which has led to the advanced research and developments in this field. In order to perform a real-time evaluation and analysis of an existing IoT network, we need large-scale and multidisciplinary testbeds, which can help us overcome the issues faced by these networks. These testbeds enable us to evaluate whether the new IoT solutions are reasonable. These testbeds also measure the degree to which these applications will be of use to the consumers [24].

3.1 IoT EXPERIMENTATION SETUPS

Most of the testbeds being used for IoT these days are adapted from wireless sensor networks (WSN), which were centered on performing experiments on networks that are isolated, and all the developments are to be done within them. However, IoT networks intend to combine these isolated networks and form a globally connected domain where devices with different configurations can communicate with each other. There are some requirements that need to be fulfilled by these testbeds in order to carry out IoT experimentations (Figure 3.1):

Increase in the number of nodes: While WSN-based testbeds contain lesser nodes, the IoT networks comprise thousands of nodes that mostly operate without any human intervention; thus, for IoT, we

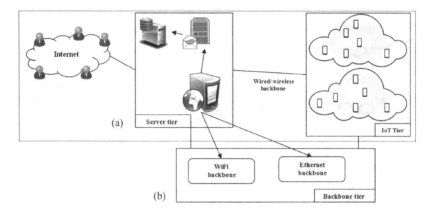

FIGURE 3.1   Design challenges for IoT testbeds (a) 2-level. (b) 3-level.

require testbeds that can perform a real-time analysis of a larger number of nodes and perform error detection and management of their own.

Heterogeneity in the devices: The testbeds should be able to support devices with different configurations. In WSNs, the nodes present at the gateway only performed the task of a sink node, whereas for IoT networks, these nodes should have other functionalities also.

Multitasking: Multitasking is a necessary feature of IoT networks. Since these systems have constraints on performance and resources, different nodes may have different tasks to carry out. The testbed supporting IoT should incorporate functionalities in order to minimize the effect of multiple experiments going on at the same time.

3.2 PLATFORMS AND SOFTWARE TOOLS FOR IoT SIMULATION

The operating systems (OS) used for IoT devices have made the integration of various software products possible, thus increasing the functionality of the network and assisting the users more in their day-to-day tasks. Although security problems in IoT-based OS are not very similar to those of the conventional ones, still the standards defined by IETF (Internet Engineering Task Force) provide protocols to secure such systems. Some of the IoT-based OS are mentioned below:

a. *mbed*: This OS was developed by Acorn RISC Machine (ARM), to be used with 32-bit Cortex-M microcontrollers. It was written in C and C++ and is an open source OS with Apache License 2.0. The mbed OS incorporates software configuration to develop firmware for IoT devices [25].

b. *RIOT*: This OS was developed by INRIA, HAW Hamburg, and FU Berlin. It is written using C and C++ and is an open source OS with LGPL v2.1. It is compatible with ARM Cortex-M3, M4, ARM7, and AVRAtmega devices. The Software Development Kit (SDK) configuration supports C and C++ applications [26].

c. *Contiki*: This OS was initially created by Adam Dunkels, but afterward, various organizations such as Cisco, SAP, and Oxford University further developed it. This OS performs efficiently for resource-constrained devices. Contiki uses Cooja (a network simulator) for Contiki nodes. There are three types of Contiki nodes: emulated nodes, Cooja nodes, and Java nodes [27].

d. *Nano-RK*: This OS is also written in C and is an open source. It was developed at Carnegie Mellon University. It is designed to work with WSN. The Eclipse IDE supports the application development for this OS [28].

3.3 INTEGRATION OF IoT WITH VARIOUS DOMAINS

3.3.1 Data Storage

The rapid growth of IoT has led to the development of new kinds of gadgets in various domains. The primary aim of any such network is the collection and processing of data in order to make decisions so as to gain knowledge about the environment. With the increase in devices, the volume of data over the Internet has also increased overwhelmingly [29].

As data is processed at various layers of IoT, it becomes more structured and readable (machine readable). Big data and cloud applications provide various tools for processing and managing the data collected by IoT sensors [30]. However, due to heterogeneity in the architecture of devices and data as well as limited processing capabilities, IoT networks have different design requirements in terms of interoperability, scalability, and security as compared to the devices currently used on the web [31].

Therefore, in order to address these requirements, Atzori et al. [2] presented a new semantic-oriented perspective for data analytics in IoT. Semantics promotes interoperability among different devices and data models and facilitates effective data management and processing and knowledge extraction. In this section, we will discuss the developments in the semantic-oriented data analytics in IoT.

a. *Semantics are not enough*: Ontologies do not make data interoperable at a global scale. Semantic annotations need to be processed and analyzed as semantic technologies are not just hype.

b. *Semantic modeling and ontology development*: Deploying semantics can simplify the process of data management as it encourages different sources with heterogeneous data to be interoperable.

c. *Linked sensor data*: Linked data enables different resources to correlate with each other. By adopting semantic annotations, IoT resources and processing can be associated with each other, which makes the scattered data more relevant. Data linking makes IoT more usable across different domains and also contributes to system interoperability. The IoT objects have semantic data, which is linked to the domain-related information of that object, for example, location. Different resources are able to link with each other due to the use of various data models and ontologies. The linked data can also be used as the source of knowledge that annotates the IoT data.

d. *Database issues in IoT*: The size and scale of data gathered by IoT devices are huge. The data collected by local IoT networks needs to be collected by a local authority. However, globally, we require a central authority to manage the domain resource knowledge using proper indexing so that looking for a certain resource or piece of data can be properly done. Some resources and information can be made publicly available, while others require access control. Most of the developments done in the IoT is service based. The SOA (service-oriented architecture) will assist the systems to integrate heterogeneous applications and ensure interoperability (Figure 3.2).

e. *Tools for data reuse*: Tools play a major role in promoting the reuse of data and ontologies. Although not many tools are currently available, still a few of them have either been newly developed or been adapted from previously existing tools to be used for IoT platforms. These tools focus on supporting widely popular ontologies and data models in order to promote common annotations and interoperability. Various tools that are used for semantic web engines can be used with IoT devices for searching and resource browsing among the linked data.

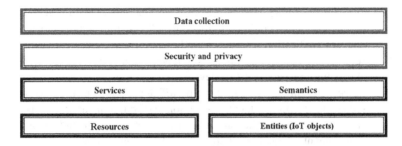

FIGURE 3.2   IoT data integration requirements.

    f. *Sense2Web-linked sensor data platform*: This tool has a GUI for data annotation, which makes it convenient for the user to link data and publish Resource Description Framework (RDF) [32]. This platform performs data linking in two ways:
- Use of globally linked resources as domain information for data annotation.
- Annotated data is published as linked data resources.

3.3.2 Cloud Computing

Cloud computing enables pervasive networks with on-demand service and a pool of shared resources. It also provides unlimited space and loses cost computation capabilities. The integration of cloud computing with IoT fulfills the requirements of IoT architecture satisfying all the necessary constraints, including application, service, and resource management [33–36]. In this section, we discuss some of the issues with the integration of IoT and cloud.

    a. *Standardization*: Most of the "things" are connected to the cloud in the current scenario resolving the storage problem and providing various services. Thus, there is an urgent need for standard protocols, interfaces, and APIs to ensure interoperability between heterogeneous platforms and facilitate interconnection among smart objects to provide enhanced services.

    b. *Integration*: The interconnection of cloud and smart objects generating and processing data in real time needs enhanced cognitive abilities and intelligent decision-making. Thus, it is necessary that there should be a common integration method with standard workflows. The combination of cloud and IoT requires a generic platform for the distribution of integrated services among different entities for managing and processing data in the integrated scenario.

c. *Scalability*: Smart devices collect huge amounts of data and thus require efficient techniques to link the gathered information with the relevant application or event. Cloud computing provides extensible support and data management, which helps in maintaining the scalability with respect to smart devices and end users.

d. *Power and energy constraints*: The IoT applications in the current scenario incorporate data transmission and communications almost at all times, thus enhancing the power consumption and energy requirements. Cloud computing provides various services such as data compression and efficient communications and caching mechanisms to decrease the power consumption and enable the reuse of cached data. Cloud also deploys middleware technologies in order to handle the long-duration data monitoring.

e. *Security and privacy*: Security and privacy in transit is always a primary concern in any application. The integration of IoT and cloud combines the real-world data and events, which makes the security threats altogether more serious. Cloud ensures properly designed policies and authorized roles so that only legitimate users have access to sensitive data (Figure 3.3).

3.3.3 Big Data

The big data produced by IoT is of different nature than the regular big data due to heterogeneity and noise in the collected data. It is estimated by HP that by 2030, the total number of sensors deployed will be in trillions, which will

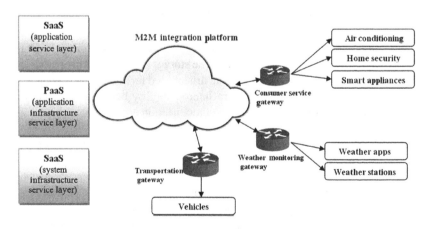

FIGURE 3.3   Integration of IoT with cloud.

make IoT a major contributor to big data [37–39]. The primary features that map IoT data to big data are as follows:

- A huge number of sensors collecting enormous amounts of data.
- Most of the gathered data is unstructured and redundant.
- Data can be transformed into useful information only after analysis.

The main features exhibited by big data produced by IoT are as follows:

a. *Large-scale data*: The data produced in IoT networks is considered to be a large scale as it requires a lot of data acquisition tools, which collect huge amounts of heterogeneous data. In order to process this data, sometimes previously processed data is required to be stored.

b. *Heterogeneity*: Data acquired by the IoT devices can be in any format and size, for example, text, audio, and video, which require a number of data acquisition devices that can gather and analyze data having different characteristics.

c. *Time and space correlation*: Time and space hold a significant priority for statistical analysis. All the data acquisition devices have location monitors and all the data packets have timestamps as the correlation between time and space is important for data analysis.

d. *Lesser amounts of effective data*: Huge amounts of data are acquired by the devices, of which only a certain amount is actually useful. For example, while acquiring traffic surveillance video, only frames that show rule violations or accidents are more significant as compared to the rest.

3.3.4 Fog Computing

The fog extends the cloud to be closer to the things that produce and act on IoT data. These devices, called fog nodes, can be deployed anywhere with a network connection: on a factory floor, on top of a power pole, alongside a railway track, in a vehicle, or on an oil rig. Any device with computing, storage, and network connectivity can be a fog node. Examples include industrial controllers, switches, routers, embedded servers, and video surveillance cameras [40].

International Data Corporation (IDC) estimates that the amount of data analyzed on devices that are physically close to the IoT is approaching 40%. There is a good reason: analyzing IoT data close to where it is collected minimizes latency. It offloads the gigabytes of network traffic from the core network, and it keeps sensitive data inside the network [41]. Analyzing IoT data close to where it is collected minimizes latency.

Fog applications are as diverse as the IoT itself. What they have in common is monitoring or analyzing real-time data from network-connected things and then initiating an action. The action can involve machine-to-machine (M2M) communications or human–machine interaction (HMI). Examples include locking a door, changing equipment settings, applying the brakes on a train, zooming a video camera, opening a valve in response to a pressure reading, creating a bar chart, or sending an alert to a technician to make a preventive repair. The possibilities are unlimited. Developers either port or write IoT applications for fog at the network edge. The fog nodes closest to the network edge ingest the data from IoT devices [42,43]. Then, this is crucial that the fog IoT application directs different types of data to the optimal place for analysis:

a. The most time-sensitive data is analyzed on the fog node closest to the things generating the data. In a Cisco Smart Grid distribution network, for example, the most time-sensitive requirement is to verify that protection and control loops are operating properly. Therefore, the fog nodes closest to the grid sensors can look for signs of problems and then prevent them by sending control commands to actuators.

b. Data that can wait seconds or minutes for action is passed along to an aggregation node for analysis and action. In the Smart Grid example, each substation might have its own aggregation node that reports the operational status of each downstream feeder and lateral.

c. Data that is less time sensitive is sent to the cloud for historical analysis, big data analytics, and long-term storage (see sidebar). For example, each of thousands or hundreds of thousands of fog nodes might send periodic summaries of grid data to the cloud for historical analysis and storage.

3.4 IoT AND RADIO FREQUENCY IDENTIFICATION (RFID)

Radio frequency identification (RFID) tags are available in many sizes, capabilities, and shapes. Which tag would be deployed into the system depends upon the business and technological requirements of the organization. The antenna, integrated circuit, printed circuit board, or substrates are the main components of RFID tags. The antenna initiates a transmission and receiving of radio wave signals for communication. It is also a coupling mechanism where energy can be transmitted in the form of electromagnetic radiation. This is the most typical style of communication between reader and tag. Depending upon the distance between antenna and reader, the antenna can also power up other components of a tag. The brain of the tag is called an integrated circuit, which is a collection of various components. Its functions are equivalent to those of a multiprocessor found in any computer or smartphone. The sole function of an integrated circuit is the transmission of a unique identifier of the tag. Moreover, it can also collect information if a tag has any peripherals and their pieces of information are also sent along with a unique identifier.

The printed circuit board is used to accommodate the tag. It may be flexible or rigid and is made up of various materials depending upon the functionality, deployment environment, purpose, and size of the tags. Tags are made with appropriate standards called a class. Six classes have been defined by EPCglobal so far as under [44]:

Class 0/class 1: Provides very fundamental radio frequency passive capabilities. The difference between these two classes is programming: class 0 is default factory-programmed and class 1 is not factory-programmed as users can program according to their own specifications.

Class 2: Possesses capabilities of encryption and RF memory read–write.

Class 3: Contains on-board batteries for longer power and also provides long-range broadband communications.

Class 4: Consists of active tags, peer-to-peer communication, and various sensors.

Class 5: Contains readers only and also activates other tags in the environment.

No built-in power sources are added into passive tags. Readers use radio frequency to power up readers and normally fall into class 0 to 3 range. Class 4 states active tags only that contain their own internal power source to provide energy for a specific time period for various operations. Class 5 deals with active tags and readers only that can read data from various tags.

RFID is one of the primary enabling technologies for IoT. RFID is replacing barcode in most of the applications. Using RFID in integration with IoT, we can add authentication mechanisms to enhance the security of devices. In the past few years, various mutual authentication schemes have been proposed for RFID, which can secure IoT devices from security threats.

IoT mainly reckons on the capability of RFID to identify and be identified by other objects and communicate with them as well. Most of the IoT applications nowadays rely on sending, receiving, and storing information. The major development issues faced by IoT are as follows:

- *Scalability*: More and more devices join the IoT network every day, which raises challenges such as addressing, services, and data management.
- *Heterogeneity*: IoT networks consist of different types of devices gathering and storing different types of data, which needs to be processed into meaningful information.
- *Power and resource constraints*: IoT devices have small memory and low power supply; thus, we need to minimize the use of power during communications.
- *Integration with WSN node*: The integration of RFID and WSN can be possible for high-end applications, but it brings a lot of challenges alongside; such types of nodes could have more than one sensor and extra communication capabilities as compared to RFID.
- *Vulnerabilities in RFID system*: The RFID reader acts as a medium between the tags and the server, and if in such cases, the number of tags increases, then the system must always be scalable. For every tag request, the server has to perform a linear search to identify the tag and the time of this search process increases with the number of tags, and it affects the performance of an authentication protocol and increases the computational overload.

Industrial Internet of Things (IIoT)

4

Current industrial trends and initiatives aim to "connect the unconnected." Today, millions of embedded devices are used in safety and security critical applications such as industrial control systems, modern vehicles, and critical infrastructure. In the past decades, classical production engineering, automation, and intelligent computation systems merged into the Industrial Internet of Things (IIoT). The number of computation components integrated into industrial control systems, production systems, and factories is steadily increasing.

With the integration of classical computing into production systems, emerging megatrends, such as mobile computing, cloud computing, and big data, are becoming important drivers of innovation in the industry. Cloud-based services are used to monitor and optimize complex supply chains. Big data algorithms predict machine failures, which reduces downtimes and maintenance costs; the interconnected production systems enable a tight integration and optimization of production and business processes as well as outsourcing production steps to other locations, companies, and freelancers. In the near future, cloud-based services will allow considering more customer requirements in the production process and planning, and enabling a new level of product individualization at a minimal cost. This development driven by computation systems is also called the "fourth industrial revolution."

Industrial IoT brings many new challenges with regard to different aspects, including security, privacy, standardization, legal, and social aspects. In particular, increased diversity and large numbers of devices in IoT systems require highly scalable solutions for, for example, naming and addressing, data communication, knowledge management, and service provisioning. Further, most IoT devices have only limited resources, which demands architectures supporting low-power, low-cost, fully networked integrated devices that are compatible with standard communication techniques.

Devices in the Internet of Things (IoT) generate, process, and exchange vast amounts of security and safety-critical data, as well as privacy-sensitive information, and hence are appealing targets of various attacks. To ensure the correct and safe operation of IoT systems, it is crucial to assure the integrity of the underlying devices, in particular their code and data, against malicious modifications. Recent studies have revealed many security vulnerabilities in embedded devices, which pose new challenges on the design and implementation of secure embedded systems that typically must provide multiple functions, security features, and real-time guarantees at a minimal cost. In this chapter, we will give an overview of the developments and trends of IIoT systems, point out related security and privacy risks and challenges, and discuss potential solutions towards a holistic security framework for IIoT systems.

4.1 M2M TO IoT

M2M, or machine-to-machine, is a direct communication between devices using wired or wireless communication channels. M2M refers to the interaction of two or more devices/machines that are connected with each other. These devices capture data and share with other connected devices, creating an intelligent network of things or systems. Devices could be sensors, actuators, embedded systems, or other connected elements [45] (Figure 4.1).

M2M technology could be present in our homes, offices, shopping malls, and other places. Controlling electrical appliances like bulbs and fans using

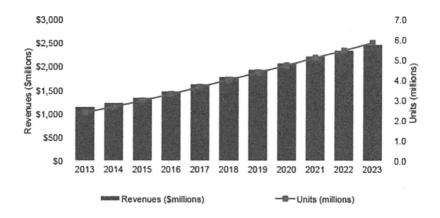

FIGURE 4.1   M2M vs IoT market revenue.

RF or Bluetooth from your smartphone is a simple example of M2M applications at home. Here, the electrical appliance and your smartphone are the two machines interacting with each other. The IoT is the network of physical devices embedded with sensors, software, and electronics, enabling these devices to communicate with each other and exchange data over a computer network. The things in the IoT refer to hardware devices uniquely identifiable through a network platform within the Internet infrastructure.

Machine-to-machine communication, or M2M, is exactly as it sounds: two machines "communicating," or exchanging, data, without human interfacing or interaction. This includes serial connection, powerline connection (PLC), or wireless communications in the IIoT. Switching over to wireless has made M2M communication much easier and enabled more applications to be connected. In general, when someone says M2M communication, they often are referring to cellular communication for embedded devices. Examples of M2M communication in this case would be vending machines sending out inventory information or ATMs getting authorization to dispense cash [46].

As businesses have realized the value of M2M, it has taken on a new name: the IoT. IoT and M2M have similar promises: to fundamentally change the way the world operates. Just like IoT, M2M allows virtually any sensor to communicate, which opens up the possibility of systems monitoring themselves and automatically responding to changes in the environment, with a much-reduced need for human involvement. M2M and IoT are almost synonymous. The exception is that IoT (the newer term) typically refers to wireless communications, whereas M2M can refer to any two machines: wired or wireless communications.

Growth in the M2M and IoT markets has been growing rapidly, and according to many reports, growth will continue. Strategy analytics believes that low-power, wide-area network (LPWAN) connections will grow from 11 million in 2014 to 5 billion in 2022. And IDC says that the market for worldwide IoT solutions will go from $1.9 trillion in 2013 to $7.1 trillion in 2020. Many big cell operators, like AT&T and Verizon, see this potential and are rolling out their own M2M platforms. Intel, PTC, and Wipro are all marketing heavily in M2M and working to take advantage of their major industry growth spurt. But there is still a great opportunity for new technology companies to engage in highly automated solutions to help streamline processes in nearly any type of industry [47,48].

However, as the cost of M2M communication continues to decrease, companies must determine how they will create a value for businesses and customers. In our mind, the opportunity and the value for M2M don't lie in the more traditional layers of the communication world. Cell carriers and hardware manufacturers, for example, are beginning to look into full-stack offerings that enable M2M and IoT product development. We strongly believe value lies in the application side of things, and the growth in this industry will be driven by smart applications from this point forward.

4.2 IoT AND SECURE AUTOMATION

The very nature of IoT devices is to collect, process, and relay data through a communication channel and sometimes control a much larger unit. The data in question can range from a heartbeat, to the temperature of a room, to living habits, and even to the location of the user. The ensuing privacy and security concerns that arise are known to manufacturers; however, security in IoT devices is either neglected or treated as an afterthought. Often, this is due to short time to market (TTM) and reduction of costs driving the device's design and development. As such, IoT devices remain open for attack and become attractive targets for those wishing to obtain the information they hold [49].

Further, given the always-on network connectivity some of these devices hold and their usage patterns, these devices could be targeted by malware, increasing the potential for harmful usage. The few devices that do choose to add any protection usually employ software-level solutions, such as firmware signing and the execution of signed binaries, methods that resemble those used in regular computing. These solutions, however, do not consider the different usage patterns that IoT devices have when compared to traditional embedded systems or personal computers. This proves to be insufficient, as not only have these mechanisms been proven to be by-passable, but software-level protection schemes often leave the hardware unintentionally vulnerable, allowing for new attack vectors [50].

In order to better understand the security and privacy issues associated with the current IoT device design flow and their implications, we used two IoT devices as case studies. Given that IoT devices are widely used in both commercial and industrial applications, the selected sample IoT devices include a smart controller for a home automation system and a smart meter for modern power grids. Commercial IoT devices that directly target end users are often designed with an emphasis on device functionality. Security features are often added in an ad hoc manner where remote attacks are treated as the main threats. Therefore, commercial IoT devices often suffer from hardware-level vulnerabilities [14], which may be remotely exploited. Similar to commercial IoT devices, smart devices are also widely used in industrial applications. These devices, if compromised, may have a more serious impact than compromised commercial IoT devices.

Improperly secured IoT devices may easily be compromised during transport and deployment by remote attackers. The effects of these attacks could range from simple backdoors to a total compromise of the device. Industrial devices pose an even bigger threat if compromised, as critical infrastructure may be damaged. In our proof of concept, we were able to change the identification of a smart meter, thus broadcasting energy consumption while masquerading as a different device. In doing this operation, an attacker is capable of altering how billing is performed for their power consumption.

Furthermore, as the meter exposes programming and debugs interfaces, it can be modified to report reduced reads, further exacerbating the problem. This has a detrimental effect on the economy, as the energy supplier is no longer able to bill customers as it should; the power grid, as expected capacity computations based on the recorded values are now incorrect; and the environment, as energy can now be used without proper recording. On-field deployment of compromised IoT devices will lead to safety complications. Due to the services these units provide, an attacker can utilize these devices to cause physical harm to users. Excess unmonitored power consumption can result in the overload of the grid, causing outages and in extreme cases, equipment failure.

By 2020, it is expected that the global smart home market will reach around 40 billion dollars. Now, these smart home devices range from anything like smart kettles, refrigerators, and dryers to air conditioners and a range of safety and security devices, like alarm systems and circuit security cameras. Ease and convenience are what make smart home systems so appealing, and as they are connected with each other, it becomes easy to manage more operations. With the help of IoT smart home devices, it becomes easy to reduce energy and costs, all the while saving time. One of the main issues that common people, as well as businessmen, are facing when it comes to applying IoT smart homes is the high costs. They are quite high compared to the nonconnected devices, and so, when it comes down to choosing IoT-enabled devices, they are always a bit hesitant.

4.3 IIoT APPLICATIONS

With the rapid growth of IoT, various applications such as in the field of healthcare and monitoring, environment and cattle monitoring, transports, and home support and security have also been growing. In this section, we discuss the applications of IoT with respect to industries. The real-time industry-based IoT applications have multiple goals. Thus, the developers need to be able to attain a balance.

However, the use of IoT is rapidly evolving and growing. Quite a few IoT applications are being developed and/or deployed in various industries, including environmental monitoring; healthcare service; inventory and production management; food supply chain (FSC); transportation; workplace; and home support, security, and surveillance. Atzori et al. [2] provide a general introduction to IoT applications in various domains. Different from their discussions, we specifically focus on industrial IoT applications. The design of industrial IoT applications needs to consider multiple goals. Depending on the intended industrial applications, designers may have to make a tradeoff among these goals to achieve a balance of cost and benefits. Below are some IoT applications in industries.

a. *IoT applications in the healthcare industry*: IoT provides new opportunities to improve healthcare. All objects in the healthcare systems (people, equipment, medicine, etc.) can be tracked and monitored constantly by being powered by IoT's ubiquitous identification, sensing, and communication capacities. All the healthcare-related information (logistics, diagnosis, therapy, recovery, medication, management, finance, and even daily activity) can be collected, managed, and shared efficiently. For example, a patient's heart rate can be collected by sensors from time to time, and then, it is sent to the doctor's office. Using the personal computing devices (laptop, mobile phone, tablet, etc.) and mobile Internet access (Wi-Fi, 3G, LTE, etc.), the IoT-based healthcare services can be mobile and personalized [51]. The widespread use of mobile Internet service has expedited the development of the IoT-powered in-home healthcare (IHH) services. Security and privacy concerns are two major challenges.

b. *IoT applications in FSC*: Today, FSC is extremely distributed and complex. It has large geographical and temporal scales, complex operation processes, and a large number of stakeholders. A typical IoT solution for FSC (the so-called Food–IoT) comprises three parts: (i) the field devices such as WSN nodes, radio frequency identification (RFID) readers/tags, and user interface terminals; (ii) the backbone system such as databases, servers, and many kinds of terminals connected by distributed computer networks; and (iii) the communication infrastructures such as WLAN, cellular, satellite, powerline, and Ethernet. As the IoT system offers ubiquitous networking capacity, all of these elements can be distributed throughout the entire FSC. Furthermore, it also offers effective sensing functionalities to track and monitor the process of food production. The vast amount of raw data can be further mined and analyzed to improve the business process and support decision-making. Big data analytics can be used to respond to the challenge of analyzing the tremendous amount of data collected from FSC.

c. *IoT applications in mining industries*: Mine safety is a big concern for many countries due to the working condition in the underground mines. To prevent and reduce any accidents in the mining, there is a need to use IoT technologies to sense mine disaster signals in order to make early warning, disaster forecasting, and safety improvement of underground production possible [19]. Using RFID, Wi-Fi, and other wireless communications technology and devices to enable effective communication between surface and underground, mining companies can track the location of underground miners and analyze critical safety data collected from sensors to enhance safety measures.

d. *IoT applications in transportation and logistics*: IoT will play an increasingly important role in the transportation and logistics industries. Furthermore, IoT is expected to offer promising solutions to transform transportation systems and automobile services. Recently, an intelligent informatics system (iDrive system) developed by BMW used various sensors and tags to monitor the environment, such as tracking the vehicle location and the road condition to provide driving directions. Zhang et al. [52] designed an intelligent monitoring system to monitor temperature/humidity inside refrigerator trucks by RFID tags, sensors, and wireless communication technology. Security and privacy protections are important for the widespread use of IoT in the transportation and logistics, since many vehicle drivers are worried about information leak and privacy invasion. Reasonable efforts in technology, law, and regulation are needed to prevent unauthorized access to, or disclosure of, the privacy data.

Recently, Ji and Qi [53] illustrate an infrastructure of IoT applications used for emergency management in China. Their IoT application infrastructure contains sensing layer, transmission layer, supporting layer, platform layer, and application layer. Their IoT infrastructure has been designed to integrate both local-based and sector-specific emergency systems. Establishing standards for implementing fire IoT is a pressing challenge now.

4.4 IIoT AND CYBERSECURITY

Security for connected consumer devices gets all the attention. However, electronics companies should also focus deeply on security for industrial systems used to manufacture components and increasingly high-tech products. The production of "intelligent industrial things" must also have effective cybersecurity, or it can place a company's entire ecosystem at risk. Our research found that more than 80% of electronics companies are implementing IIoT technologies in plants and assembly lines without fully evaluating the risks or preparing effective responses.

Electronics companies need cybersecurity capabilities that are contextual, cognitive, and adaptive to continuously identify, mitigate, and prevent risk. Manufacturing operations are among the most expensive parts of the electronics value chain. While IIoT provides insight, it can also increase the risk of exposure to potential cyberattacks and damage on multiple fronts. Each point presents a new opportunity for unauthorized entry. Whether caused by cyber hackers, competing companies, countries engaged in corporate espionage, or even disgruntled employees, losses can mount quickly once under attack. The risks may include equipment failure, loss of critical data and corporate reputation, or even injury and loss of life (Figure 4.2).

IIoT technologies can vastly improve operational efficiencies, yet they also expose potential new attack surfaces and security targets if not properly protected. Each new machine joins "a system of systems" as it connects to additional IIoT devices. Technological expansions such as 5G will likely increase the use of IIoT technologies by providing the infrastructure needed to

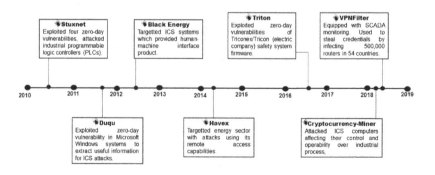

FIGURE 4.2 Attacks on industrial control systems.

carry huge amounts of data. But this also widens the attack surface. Virtually anything can become vulnerable, from high-value assets or services, critical workloads in the cloud, process control subsystems in cyber-physical systems to critical business and operational data.

Organizations need capabilities that will protect not only their assets and networks, but also their entire IIoT ecosystems. Equally important is the ability to respond quickly and effectively in the event of a breach. Organizations of virtually all types must work to keep pace with ever-evolving IIoT threats. To better understand IIoT security risks and implications, the IBM Institute for Business Value (IBV) partnered with Oxford Economics to survey 700 executives. They represent 700 companies in 18 countries from the energy and industrial sectors, of which 269 were electronics. They are all implementing IIoT in their plants. The two most prevalent applications are for real-time equipment monitoring and predictive maintenance, at 65% and 58%, respectively. Automation of machines and processes are also common applications, with 45% and 43% using IIoT technologies to automate machines and workflow, respectively.

Electronics companies are aware of the cybersecurity risks and are working to manage their security spending accordingly. But they are less clear on the combination of IIoT cybersecurity capabilities – skills, controls, practices, and protective technologies – required to secure their current and future business from IIoT threats.

4.5 CONCLUSION

To be cyber resilient, IoT manufacturers need to spend more time on security measures in the product development stage. Products that come with reliable endpoint protection and detailed monitoring infrastructure are well placed to meet the security challenges of IIoT in 2018. Another primary task for an IIoT deployment team is meeting encryption standards. Every interaction with the system (from any device) must go through generally accepted cryptography protocols. The volume of personal information being shared every day across the network will only rise in the years to come – making encryption one of the most critical measures from the perspective of privacy. As more devices connect to a shared network, it becomes harder for IT teams to identify risks. As the WEF suggest, IIoT deployment teams must build modules that define all the digital and physical assets that need protection. They also need to better collect data on potential threats. One of the most significant problems is when integrity legacy software with IIoT platforms is a lack of maintenance. While testing legacy software is one challenge, it is imperative that IIoT devices (and associated software) are tested rigorously throughout their lifespan. Methodologies should include "unit, system, acceptance, regression testing, and threat modeling, along with maintaining an inventory of the source for any third-party/open source code and components utilized."

Trust and Privacy in IoT

5

The Internet of Things (IoT) presents numerous benefits to consumers and has the potential to change the ways that consumers interact with technology in fundamental ways. In the future, the IoT is likely to meld the virtual and physical worlds together in ways that are currently difficult to comprehend. From the perspective of security and privacy, the predicted pervasive introduction of sensors and devices into currently intimate spaces – such as the home, the car, and with wearables and ingestible, even the body – poses particular challenges. As physical objects in our everyday lives increasingly detect and share observations about us, consumers will likely continue to want privacy.

The IoT is emerging as the third wave in the development of the Internet. The 1990s' Internet wave connected 1 billion users, while the 2000s' mobile wave connected another 2 billion. The IoT has the potential to connect 10× as many (28 billion) "things" to the Internet by 2020, ranging from bracelets to cars. Breakthroughs in the cost of sensors, processing power, and bandwidth to connect devices are enabling ubiquitous connections right now. Smart products like smart watches and thermostats (Nest) are already gaining traction as stated in Goldman Sachs Global Investment Research's report. The rise of the IoT has already sparked concerns about privacy: now security pros are worried that badly configured gadgets might provide a backdoor for hackers looking to break into corporate networks.

5.1 PRIVACY IN IoT

Protecting consumer privacy becomes increasingly difficult as the IoT becomes more prevalent. More devices are connected to different types of devices, and this increase in connectivity and data collection results in less control. Both control of data and control of the very devices that are connected are at stake.

Control can be lost if someone hacks into the smartphone or computer acting as a remote for the other devices. In the case of computers and smartphones, this hacking can be done remotely and often undetected. Smartphones, just like computers, carry an enormous amount of personal information about their owners. They often link to bank accounts, email accounts, and in some cases household appliances. Stolen data can result in serious problems. Vehicles contain many computers that control their function. Initially, these computers could not be hacked into. With the increased connectivity of the IoT, however, vehicles are now at risk due to being connected to the Internet.

In another sense, control can be lost as more and more companies collect data about users. This data often paints a detailed picture of individual users through the collection of activities online. Everything we search, all of the activities online, is being tracked by companies that use that data. These companies often use the data to improve the user's experience, but they also use this data to sell user's products or sell to other companies who sell user's products. Innovation in this realm means that the organization must alter the privacy policies that are in place as well as how they interact with these devices. Companies will need to take another look at the policies that they have in place to ensure that consumers are offered opportunities to access and control their own data. Consumers will become increasingly aware of the privacy implications of this level of connectivity through an interaction with the IoT and exposure to the policies that companies provide to them.

Frank Pasquale, law professor and EPIC advisory board member, discusses privacy concerns related to the IoT in a May 2014 Pew Research Report. He states that the expansion of the IoT will result in a world that is more "prison-like" with a "small class of 'watchers' and a much larger class of the experimented upon, the watched." In another article, he reinforces the idea that the IoT "will be a tool for other people to keep tabs on what the populace is doing" [54].

EPIC President, Marc Rotenberg, explains in the Pew Research Report that the problem with the IoT is that "users are just another category of things," and states that this "is worth thinking about more deeply about in the future." This is where the concept of "privacy by design" comes into play. In this approach, manufacturers assess a product's potential privacy risks and considerations during the design phase, and then adjust or address those issues in the product's development and manufacturing process [55]. This concept is a key feature in the General Data Protection Regulation (GDPR), in which the most privacy-friendly settings such as those that collect, retain, and share personal information will be required to be designed and built into new products, devices, and business processes when the rule takes an effect in 2018.

In general, a manufacturer's IoT device privacy considerations should reflect every step of the planned operating life and retirement of a device, including how to address any associated cloud services, how much data really needs to be gathered, and what should be done with any retained data. For instance, these considerations could address the secondary market for long-lived items such as refrigerators or cars.

5.2 THREAT TO DATA PRIVACY IN IoT

Above all regulatory threats, privacy will loom the largest for many IoT systems and services because it is a form of regulation that cuts across many industries and IoT-use cases. Additionally, privacy advocates have been known to operate above technology and yet seek to propose (or impose) regulation with an incomplete or flawed understanding of systems and services. Privacy is also a high priority among legislators, but it is an inconsistently defined regulatory requirement with many different privacy laws overlapping at different levels of government. Privacy regulation can be found as omnibus legislation that applies across a national economy, but also as industry-specific guidance. For instance, health systems will often have their own privacy statutes and regulations over and above the omnibus legislation.

Many of the emerging use cases in the IoT are potentially inconsistent with personal privacy: they are founded on the assumption that personal information about users is accessible and linkable. In some cases, this is a valid assumption, especially if the service in question is free and the business model calls for selling directed advertising or user profile information to a third party. But a real threat to business and user alike will arise as expectations of privacy do not match the business designs of an IoT service provider. In sum, privacy threats in the IoT come in two broad forms [56]:

- *Type 1 privacy threat*: Threats to businesses that need to collect, use, and disclose personal data, and their ability to do so without legal or market-based sanctions for breaches.
- *Type 2 privacy threat*: Threats to individuals and users of IoT service, who will be offered services that are either free or subsidized, through consenting to the collection, use, and disclosure of personally identifiable information.

IoT finds its applications in many different fields, for example, patients' remote monitoring, energy consumption control, traffic control, smart parking system, inventory management, production chain, customization of the shopping at the supermarket, and civil protection. For all of them, users require the protection of their personal information related to their movements, habits, and interactions with other people. In a single term, their privacy should be guaranteed. In literature, there are some attempts to address such an issue.

In [57], a data tagging for managing privacy in IoT is proposed. Using techniques taken from the Information Flow Control, data representing network events can be tagged with several privacy properties; such tags allow the system to reason about the flows of data and preserve the privacy of individuals. Exploiting tagging within resource-constrained sensor nodes may not be a viable solution because tags may be too large with respect to the data size and sensitivity; therefore, they generate an excessive overhead. Clearly, in this case, it is not suitable for IoT.

In [58], a user-controlled, privacy-preserved access control protocol is proposed, based on context-aware k-anonymity privacy policies. Note that privacy protection mechanisms are investigated: users can control which of their personal data is being collected and accessed, who is collecting and accessing such data, and when this happens.

In [59], it is presented Continuously Anonymizing Streaming data via adaptive clustering (CASTLE). It is a cluster-based scheme that ensures anonymity, freshness, and delay constraints on data streams, thus enhancing those privacy-preserving techniques (e.g., k-anonymity) that are designed for static data sets and not for continuous, unbounded, and transient streams. More in detail, [58] models k-anonymity on data streams and defines k-anonymized clusters exploiting the quasi-identifier attributes of tuples in order to preserve the sensitive data privacy.

In [60], the traditional privacy mechanisms are divided into two categories: discretionary access and limited access. The former addresses the minimum privacy risks, in order to prevent the disclosure or the cloning of sensitive data, whereas the latter aims at limiting the security access to avoid malicious unauthorized attacks.

5.3 ENFORCING TRUST IN IoT

The trust concept is used in various contexts and with different meanings. Trust is a complex notion about which no definitive consensus exists in the scientific literature, although its importance is widely recognized. The main problem with many approaches towards trust definition is that they do not lend themselves to the establishment of metrics and evaluation methodologies. Moreover, the satisfaction of trust requirements is strictly related to the identity management and access control issues. Both [59,60] focus on the trust-level assessment of IoT entities. The authors assume that most smart objects are human-carried or human-related devices, so they are often exposed to public areas and communicate through wireless, and hence vulnerable to malicious attacks. Smart objects have heterogeneous features and need to cooperatively work together. The social relationships considered are friendship, ownership, and community, since users are friends among themselves (i.e., friendship), users own the devices (i.e., ownership), and the devices belong to some communities (i.e., community). Malicious nodes aim at breaking the basic functionality of IoT by means of trust-related attacks: self-promoting, bad-mouthing, and good-mouthing. The trust management protocol for IoT proposed in [61] is distributed, encounter based, and activity based: two nodes that come in touch to each other or involved in a mutual interaction can directly rate each other and exchange trust evaluation about the other nodes, so they perform an indirect rate which seems like a recommendation. The reference parameters to trust evaluation are honesty, cooperativeness, and community interest.

Therefore, such a dynamic trust management protocol is capable of adaptively adjusting the best trust parameter setting in response to dynamically changing environments in order to maximize application performance. A similar approach to provide a trustworthiness evaluation is carried out in [62] in the so-called Social Internet of Things (SIoT). This paradigm derives from the integration of social networking concepts into IoT, due to the fact that the objects belonging to the IoT infrastructure are capable of establishing social relationships in an autonomous way with respect to their owners. The challenge addressed in [63] is to build a reputation-based trust mechanism for the SIoT, which can effectively deal with certain types of malicious behaviors aimed at misleading other nodes, in order to drive the use of services and information delivery only towards trusted nodes. A subjective model for the management of trustworthiness is defined, which builds upon the solutions proposed for P2P networks. Each node computes the trustworthiness of its friends based on their own experience and the opinion of the common friends. As a consequence, a node chooses the provider of the service it needs based on this highest computed trustworthiness level.

Yet in relation to the social network context, in [64], the authors propose a secure distributed ad hoc network; it is based on a direct peer-to-peer interactions and the creation of communities in order to grant a quick, easy, and secure access to users to surf the web; thus, it is close to the social network concept. Each node (i.e., device) and each community have an identity in the network and modify the trust of other nodes based on their behavior, thus establishing a trust chain among users. The parameters analyzed are physical proximity, fulfillment, consistency of answer, hierarchy on the trusted chain, similar properties (e.g., age, gender, types of sensors), common goals and warrants, history of interaction, availability, and interactions. Chains of confidence will allow the establishment of groups or communities and unique identities (for the communities) for the access to services as well as for the spreading of group information. Therefore, security is established when the users access the network through the use of the trust chain generated by nodes, which he/she crosses. It is considered that the traditional access control models are not suitable for the decentralized and dynamic IoT scenarios, where identities are not known in advance. Trust relationship between two devices helps in influencing the future behaviors of their interactions.

When devices trust each other, they prefer to share services and resources. Such a paper presents a Fuzzy approach to the Trust-Based Access Control (FTBAC). The trust scores are calculated by the FTBAC framework from factors such as experience, knowledge, and recommendation. Such trust scores are then mapped to permissions, and access requests are accompanied by a set of credentials, which together constitute a proof for allowing the access or not.

5.4 TRUST MANAGEMENT

In order to provide an IoT network that is trustworthy, the following goals should be achieved [65,66]:

a. *Correlation and agreement*: Trust management enables the assessment of the degree of correlation and agreement between IoT network nodes. It also helps them to reach consensus and cooperate with each other. Trust correlation assessment takes into account all the entities present within the IoT network at all the layers and plays a fundamental role in intelligent and autonomic trust management.

b. *Knowledge apprehension*: Trust management system should ensure a reliable data transfer between nodes in the network. In order to achieve that, the system must ensure precision and perseverance as well as an efficient collection of data. This goal is generally achieved at the perception layer.

c. *Maintaining confidentiality*: Trust management system must act in accordance with policies in order to safeguard the privacy and confidentiality of sensitive data, which are in transit within the IoT network.

d. *Trustworthy communication*: Trust management system must ensure that the data flow and messages passing within the network are secure from any threats. For this purpose, authorization certificates and key management are deployed to control any malicious access in the network. This goal of trust management is one of the fundamental security requirements for IoT's development.

e. *Quality of service*: This objective of the trust management system is mainly application specific (application layer). However, it requires support from other layers also. Quality of service ensures that the services should be provided to the right end user at the right time.

5.5 CONCLUSION

The real spreading of IoT services requires customized security and privacy levels to be guaranteed. The broad overview provided with this survey arises many open issues and shed some light on research directions in the IoT security field. More in detail, a unified vision regarding the insurance of security and privacy requirements in such a heterogeneous environment, involving different technologies and communication standards, is still missing. Suitable solutions need to be designed and deployed, which are independent from the exploited platform and able to guarantee: confidentiality, access control, and privacy for users and things, trustworthiness among devices and users, compliance with defined security and privacy policies. Research efforts are also required to face the integration of IoT and communication technologies in a secure middleware, able to cope with the defined protection constraints. Another research field is that of IoT security in mobile devices, increasingly widespread today. Much efforts have been (and are being) spent by the worldwide scientific community to address the aforementioned topics, but there are still many open issues to be faced.

Authentication Mechanisms for IoT Networks

6

Developing comprehensive security and privacy solutions for IoT requires revisiting almost all security techniques we may think of. Encryption protocols need to be engineered so as to be efficient and scalable for deployment on large-scale IoT systems and devices with limited computational resources. Benchmarks are needed to perform detailed assessments of such protocols. In addition, as devices may be physically unprotected, attackers may have access to the state of the memory, while encryption operations are being performed. Addressing such problems may require new techniques based, for example, on white-box cryptography.

White-box encryption techniques hide encryption keys by transforming them into large look-up tables in order to make harder for attackers to extract the keys. Such techniques are, however, very expensive, and many of the proposed white-box encryption protocols have been cryptanalyzed. Introducing dynamics in the look-up tables by a shuffling approach may help in addressing such a problem. In addition, scalability of such protocols is critical, in that in many safety-sensitive applications, encryption operations must be very efficient. For example, in a vehicle network, a message from a vehicle informing other vehicles of a sudden break should be processed very quickly in order to give the other vehicles enough time to break.

6.1 DATA SECURITY IN IoT

The estimated growth of this new trend in the market is expected to hit between 26 billion and 30 billion devices by 2020, with an estimated market worth of between $6 trillion and $9 trillion. To put this in context, the following are some interesting implications (including ones concerning data protection) that relate to the explosion of these interconnected devices [67,68]:

- These devices will constantly generate huge amounts of data, so we will need faster networks, larger storage capabilities (likely in the cloud), and more bandwidths to support the growth in Internet traffic.
- There is not yet an open ecosystem to host these devices to make them interoperable as there is on Microsoft Windows, Apple iOS, and Google Android ecosystems.
- Vendors are creating private networks for interoperability among their own products, but these are incompatible with others. This creates a major challenge for integration across multiple solutions.
- The current Internet protocol (IPv4) cannot handle the growth in the number of interconnected devices on the Internet. This will trigger the need to switch to a more scalable protocol, such as IPv6.

Data security, availability, and quality are critical areas for IoT. Data security requires, in addition to the use of encryption to secure the data while being transmitted and at rest, access control policies to govern access to data, by taking into account information on data provenance and metadata concerning the data acquisition context, such as location and time. Availability requires among other things to make sure that relevant data is not lost. Addressing such requirement entails designing protocols for data acquisition and transmission that have data loss minimization as a key security goal. Kinesis [69] is an example of a sensor network system designed to make it possible for sensors to automatically take response actions in the event of data transmission disruptions. Ensuring data quality is a major critical requirement in IoT as data acquired and transmitted by IoT devices may be of poor quality, because of several reasons such as bad device calibration, device faults, and deliberate attacks aiming at data deception attacks. Solutions like data fusion need to be revised and extended to deal with dynamic environments and large-scale heterogeneous data sources.

6.2 IoT LIMITATIONS

Devices of the IoT are resource constrained, and therefore, traditional security mechanisms are not precise in smart things. According to [70], some security limitations related to the IoT communication devices are as follows:

- *Memory capacity*: Restricted devices use random access memory (RAM) to store data, and their storage capacity ranges between a few kilobytes and 12 kilobytes. Data storage in the IoT devices is limited, and some devices cannot store or send data. As a result, some data is ignored if it exceeds the limit of storage.
- *Energy capacity*: It is the amount of energy the devices have to maintain itself over a specified period. The energy sources in the devices are limited and need to be replaced after a particular time. Some IoT devices consume large amounts of power and are not rechargeable. Therefore, to save the battery in limited devices, low-bandwidth connections are used.
- *Processing capacity*: It refers to the amount of power in the devices. Many IoT devices are small sized and of low cost with low processing capacity. Therefore, these devices require lightweight protocols to work efficiently.

6.3 NEED FOR AUTHENTICATION AND ACCESS CONTROL IN IoT

IoT is focusing on machine-to-machine (M2M) mode of communication. For such communication nodes, authentication is very important for insuring security and privacy. When two or more nodes are communicating with each other for a common objective, they should authenticate each other first in order to block fake node attack. However, there is no efficient authentication mechanism for a massive number of IoT devices.

The first principle of security is visibility. Issues from patching to monitoring to quarantining all require establishing visibility the moment a device touches the network. Since access control technologies are usually the first network element that a new device touches, they need to be able to automatically recognize IoT devices, determine if they have been compromised, and then provide controlled access based on factors such as type of device, intended destination, and if it is user based, the role of the user.

As the volume of IoT devices connecting and disconnecting from the network continues to escalate, access control solutions need to be able to do this at digital speeds. Access control solutions then need to seamlessly synchronize with network and security controls to ensure that these devices are tracked and that policy enforcement is consistently enforced as they and their applications move across the distributed network.

Strong IoT device authentication is required to ensure that connected devices on the IoT can be trusted to be what they purport to be. Consequently, each IoT device needs a unique identity that can be authenticated when the device attempts to connect to a gateway or central server.

6.4 SYSTEM REQUIREMENTS FOR AUTHENTICATION MECHANISMS

The main IoT security concerns are authentication, authorization, integrity, confidentiality, non-repudiation, availability, and privacy.

- *Authentication*: The process of confirming and insuring the identity of objects. In IoT context, each object should have the ability to identify and authenticate all other objects in the system (or in a given part of the system with which it interacts).
- *The authorization*: The process of giving permission to an entity to do or have something.
- *Integrity*: The way towards keeping up the consistency, precision, and dependability of information over its whole life cycle. In IoT, the alteration of basic information or even the infusion of invalid information could prompt major issues; for example, in smart health system-use cases, it could lead to the death of the patient.
- *Confidentiality*: The process of ensuring that the information is only accessed by authorized people. Two main issues should be considered regarding confidentiality in IoT: first, to ensure that the object receiving the data is not going to move/transfer these data to other objects and, second, to consider the data management.
- *Non-repudiation*: The way towards guaranteeing the ability to demonstrate that a task or event has occurred (and by whom), with the goal that this cannot be denied later. In other words, the object cannot deny the authenticity of a specific data transferred.

- *Availability*: The process of ensuring that the service needed is available anywhere and anytime for the intended users. This includes in IoT the availability of the objects themselves.
- *Privacy*: The process of ensuring nonaccessibility to private information by public or malicious objects.

6.5 TAXONOMY OF IoT AUTHENTICATION PROTOCOLS

Depending upon the communication and computation capabilities, the mutual authentication protocols are divided into four classes (as shown in Figure 6.1): (i) full-fledged protocols, use classical cryptographic encryption and decryption and have high computational overhead; (ii) simple protocols, use lightweight techniques such as elliptic curve cryptography (ECC), pseudo-random number generator (PRNG), and hash function; (iii) lightweight protocols, use checksums, hash function, and bitwise operations; and (iv) ultra-lightweight protocols, only use bitwise operations, for example, rotations and permutations. There are several mutual authentication protocols for radio frequency identification (RFID) tags, which can be used to secure IoT devices. Some of these protocols are given in Table 6.1.

6.5.1 Hash-Based Protocols

Hash-chain based approaches deploy cryptographic mechanisms to secure passwords and other credentials. A hash chain of length "N" is obtained by recursive application of one-way hash function with an initial value "x": $h^N(x) = h(h(\dots h(x)\dots))$, such that if an adversary knows $h^N(x)$ and "x," it cannot obtain $h^{N-1}(x)$.

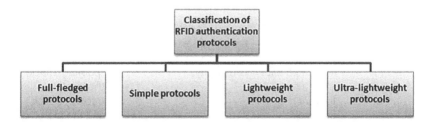

FIGURE 6.1   Classification of RFID authentication protocols.

TABLE 6.1 RFID Security Techniques

	DESCRIPTION	NOVELTY	SHORTCOMINGS
Chen et al. [71]	Private mutual authentication scheme based on quadratic residues referred to Rabin cryptosystem	Server performs only one decryption to obtain the real identify of tags without requiring linear search	Suffers from forgery attack, location privacy problem, and replay attack.
Yeh et al. [72]	Securing RFID systems conforming to EPC Class 1 Generation 2 standard	Able to detect DoS attacks and also enhances the overall performance efficiently at the same time	Cannot withstand against server impersonation attack and the data integrity attack since the server's data is transmitted in the plaintext format
Cho et al. [73]	Hash-based RFID mutual authentication protocol using a secret value	Enable a constant creation of distinct response messages without interferences from intended or meaningless requests generated by an adversary, while the secret value is not directly transmitted	Cannot withstand against resynchronization attack, the tag impersonation attack, and the reader impersonation attack
Burmester-Munilla et al.'s protocol [74]	Tag and server are mutually authenticated by exchanging either three or five consecutive numbers from the PRNG	Supports session unlinkability and forward and backward security. In this protocol, each tag shares with the server a synchronized PRNG.	Performance issues as the PRNG needs to be refreshed when there is suspicion that the state of the PRNG may be compromised
Liao and Hsiao's protocol [75]	ECC-based RFID authentication scheme integrated with ID-verifier transfer protocol	Supports forward and backward security and provides very low time complexity.	Vulnerable to the tag masquerade attack, the server spoofing attack, the location tracking attack, and the tag cloning attack

(Continued)

TABLE 6.1 (*CONTINUED*)   RFID Security Techniques

	DESCRIPTION	NOVELTY	SHORTCOMINGS
Chou's protocol [76]	ECC-based RFID authentication scheme using prime and random numbers	Low computation complexity and all the messages need to be communicated after using hash function as the channel is assumed to be insecure.	Suffers from the tag information privacy problem, backward traceability and forward traceability problem
Tu and Piramuthu's protocol [77]	Authentication protocol for session initiation protocol using smart card.	Password-based key agreement; thus, the user is free to change their password in case of any suspicion	Vulnerable to impersonation attack

The first hash-based approach was proposed by Ohkubo et al. [78], which used an identifier that was new for every session. This protocol did not use any random number generator. The parameters are renewed for every new session. They used a fresh tag identifier for every new session without using any random number generator. However, the protocol is vulnerable to replay attacks.

Weis et al. [79] proposed one-way hash function-based security scheme. Each tag has a temporary ID, which is the result of hash of a key "k." Each tag functions in either locked or unlocked state. During the lock state, the tag responds to all the requests with the temporary ID and no other action is taken.

Henrici and Muller [80] proposed an authentication protocol using one-way hash functions to ensure secure communication between the reader and the server. After the successful completion of every session, the tag ID is updated at both server's and tag's sides in order to ensure location privacy and forward secrecy. The protocol also ensures anonymity; however, it fails to provide security against backtracking.

Molnar and Wagner [81] proposed another authentication protocol that used a hash tree approach instead of hash chains to distribute the cost of computations more evenly throughout the RFID system (among all the nodes). They also developed a new model library to store node details at the backend server so as to prevent exhaustive search and ensure more security. The library architecture was very efficient in terms of searching speed; however, the authentication protocol failed to ensure anonymity.

6.5.2 ECC-Based Protocols

The current state-of-the-art authentication protocols for RFID systems include various categories of solutions such as mechanisms built up on public key cryptography or other mechanism to provide security at low cost by using simpler techniques to ensure tag security. The use of public key cryptography for low-cost devices is not a very efficient idea as it requires a lot of resources and space. Similarly, if we attempt to use existing symmetric key solution with these devices, the protocols fail to work efficiently due to different network conditions and lesser resources.

ECC has been providing very efficient substitution to public key cryptography for low-cost devices because it requires very small keys to provide security and privacy comparable to traditional public key cryptography-based schemes. The first ECC-based authentication protocol for RFID devices was proposed by Tulys and Batina in 2006 [82]. However, later on, Lee et al. [83] found out some weaknesses in this protocol and proposed a new solution to address these issues.

Liao and Hsiao [84] proposed another ECC-based solution that did not use any hash functions or complex computations, and hence, it was very simple to implement and was suitable for passive tags also. Later on, Peeters and Hermans [85] proved that the aforementioned protocol is unable to ensure security against tracking, spoofing, and cloning. It also fails to ensure privacy. Liao and Hsiao [75] proposed another authentication protocol, which claimed to ensure privacy and anonymity; they also presented proofs that the protocol ensures security against tracking and tag cloning.

Tan [86] proposed a three-factor-based key exchange mechanism for secure communication between low-cost devices. However, this scheme was unable to secure the devices against Denial of Service (DoS) and replay attacks. To overcome the vulnerabilities of Tan's protocol [86], Arshad and Nikooghadam [87] presented an ECC-based authentication protocol. However, Lu et al. [88] showed that Arshad and Nikooghadam's [87] protocol is vulnerable to password attacks, which can make tags vulnerable to impersonation.

6.5.3 PUF-Based Protocols

A physically unclonable function (PUF) is used to map a set of challenge values to set of response values based on a complex function. The complex function is based on a physical system and computes a response value for each challenge. PUFs provide a cost-effective security protocol. There are various ways to implement a PUF, for example, using silicon physical systems, which hides timing and delay of Integrated Circuits (ICs). PUFs are embedded in devices and track even various changes in the host environment, for example, temperature and pressure. A different response value will be generated every time the host environment changes. Thus, PUFs are hard to predict due to different responses even with a slight change in the challenge values. However, some solutions [89–93] suggested that in order to obtain the secret value from the token, the PUF has to deploy some error correction techniques, which increase the cost of computations.

Some other works have also explored various weaknesses that exist in PUF-based functions. Most of the PUFs fail to ensure a mutual authentication between the tag and the reader. PUF-based approaches also have scalability issues and are prone to DoS attacks, which can disrupt the connection and services between the server and the tag permanently.

6.5.4 HB Protocols

Hopper and Blum [94] proposed an authentication scheme in 2001, known as HB protocol. These protocols rely on Learning Parity with Noise (LPN) problem, which is a computationally hard problem.

LPN problem [95]: Assume P to be a matrix with the dimension q, x to be a k-bit binary vector, and v to be a vector such that $wt(v) \leq q \cdot \eta$, where $\eta \in [0, \frac{1}{2}]$ and the size of v is q-bits. If we have $z = P \cdot x \oplus v$, then the LPN problem is to obtain x_0 such that $|P \cdot x_0 \oplus v| \leq q \cdot \eta$.

LPN has various applications in cryptography as it is able to provide a stepping stone for provably secure solutions; that is, it can be mathematically shown that the solution is able to resist an active attacker in the existing environment. These schemes are very simple and take only a few step; thus, these schemes are very feasible for low-cost devices. The cryptographic schemes are mostly decisional; for the LPN problem, it can be shown that there is not a very significant difference between the decisional approach and the search approach. The cryptographic notions that can be derived from this problem can be collision resistant. The LPN problem is assumed to be collision resistant and remains a hard problem even for nonuniform noise values.

Juels and Weis [96,110] presented another HB-based approach for RFID systems referred to as HB+ [97]. They also demonstrated an active attack and added some other minor computations to improve the HB protocol. Later on, Gilbert et al. [98] proved that both HB and HB+ schemes are vulnerable to man-in-the-middle attacks.

6.5.5 Ultra-Lightweight Protocols

The fourth class of protocols is the ultra-lightweight protocols, which are entirely developed using bitwise operations such as OR, AND, XOR, rotation, or permutation. These protocols have the lowest overhead in terms of storage and computation.

In SASI [99], each tag has an ID and shares a pseudonym (IDS) and key value with the backend database server. The length of each of them is 96-bits. SASI ensures strong authentication and integrity, and uses bitwise XOR ($\oplus$)($\oplus$), bitwise OR ($\vee$)($\vee$), bitwise AND ($\wedge$)($\wedge$), addition mod $2n$(+), and left rotate ($Rot(x, y)$) operation, which left rotates the value of x with y-bits. Complex operations such as hash functions are not used by this protocol. However, this protocol is susceptible to disclosure attacks and does not ensure untraceability.

Peris-Lopez et al. proposed LMAP [100] protocol that used simple bitwise operations XOR ($\oplus$)($\oplus$), bitwise OR ($\vee$)($\vee$), bitwise AND ($\wedge$)($\wedge$), and addition mod 2m (+). This protocol ensures a mutual authentication and security from various attacks without the use of complex operations like hashing. This scheme uses an index pseudonym (IDS), which is 96-bits in length. Here, the IDS is the index of the row where all the tag-related data are stored. Each tag has key, which is divided into four parts of 96-bits each.

M2AP [101] protocol, which is very similar to LMAP [100], is also a lightweight mutual authentication protocol for RFID tags, where the index pseudonym updation procedure is different from LMAP while key updating operations remain the same. Both LMAP and M2AP [101] ensure anonymity and mutual authentication and provide security against various attacks, such as replay attacks and man-in-the-middle attacks. However, both of these protocols are susceptible to desynchronization and full-disclosure attacks.

Another protocol called EMAP [102], which is based on challenge–response mechanism, is an authentication scheme for passive tags. Most of the complex computations in this protocol are performed by the reader, and tags perform lightweight operations such as hash. It only requires one storage unit for the tag in addition to the ID for storing authentication-related data. This protocol also ensures confidentiality, integrity, and untraceability.

Peris-Lopez et al. proposed the Gossamer protocol [103], which addresses the weaknesses of SASI [99] such as desynchronization and disclosure attacks. It uses dual rotations and MixBits operation, which is a lightweight function (i.e., combination of bitwise right shift and addition operations). However, this protocol has low throughput.

6.6 CONCLUSION

IoT is the next step towards using Internet anywhere and anytime. IoT allows to connect people and devices (things) anytime and anyplace, with anything and anyone. The main security issues related to IoT are explained in brief. By observation, it is easy to understand that there is no project still in work, which satisfies all security issues in IoT. Also, there is no single project that provides policy enforcement in the IoT. In summary, an accomplished vision admiring the assurance of security and privacy constraints in a dissimilar environment, which implies that current security services are insufficient for such contradictory technologies and communication standard. As IoT deals with interconnecting various heterogeneous things, currently there are many challenges occurring while building it. So this area has many open research issues. The future research directions mainly consist of how to deal with the challenges, may be related to security issues, faced by IoT.

Provable Security Models and Existing Protocols

7

Internet of Things (IoT) devices have restricted resources and energy; however, they require regular updates and identification from other devices and the backend server. Thus, security is also a major concern, and the most common way to do it is mutual authentication between the communicating parties. The authentication schemes used for RFID tags are generally lightweight considering the constraints over IoT devices.

The RFID tags have a capacity of few kbs and deploy around 5,000–6,000 gates, out of which only a few hundred can be devoted to security functions. However, the classical cryptography authentication protocols require around 30,000–40,000 logic gates. In addition to that, we also have to consider the limited battery power.

However, in spite of these issues, the use of RFID tags in various fields has been increasing rapidly. Nevertheless, we cannot ignore the privacy issues related to these systems. The items deploying these tags may reveal confidential data such as location due to predictability. A very common solution to this issue is the use of random values as keys. There are many more issues such as above, which are faced by RFID systems. For example, various security models do not take into consideration that an adversary could be able to attack a tag physically, for example, tampering of the tag.

However, this situation must always be considered as products with low cost can be physically compromised very easily. Any adversary can carry out "reset" attacks and side-channel attacks by influencing the physical conditions of the tags, that is, varying power and voltage. Some models, however, take this into consideration while examining a protocol's security.

Many researchers have been working on these past couples of years to ensure the security and privacy of RFID systems. However, there is still no generic solution to this problem as most of the existing solutions are specific to particular scenarios or provide security against certain types of attacks. In addition to finding security solutions for RFID devices, researchers are also working on developing formal models, which can be used to prove the strength of protocols. These models help in analyzing the design and working of RFID protocols. They also emphasize the strengths and weaknesses of a protocol. However, these models also suffer from various deficiencies; for example, most of the frameworks do not fully comprehend the effects of functionalities and the level of access provided to the adversary on the system.

7.1 PROVABLE SECURITY MODELS

Most of the authentication schemes these days turn out to be insecure as they are developed solely based on previous experiences. But provable security has emerged as a new state-of-the-art model for the verification and development of more secure protocols based on experience as well as cognitive reasoning. Provable security aims to provide concrete mathematical proof for ensuring the protocol's security.

Modern cryptographic protocols deploy "game-based" security models, which have an adversary (active or passive) that can carry out potential attacks over the concerned system. If there exists an adversary who can win the game, then the scheme is considered to be insecure.

7.1.1 Vaudeney's Model

This model [104] has the capability to affect all the ongoing communications between entities and is able to perform a man-in-the-middle attack on any entity accessible to the adversary. It also has the authority to get the output of a device authentication (0 or 1). The adversary in this model is able to select or deselect tags randomly which are moving in and out of the range. The adversary has the ability to corrupt tags by extracting the details of their internal state. During any ongoing session, the adversary has a temporary identifier for the tags within its range.

Oracles: the following oracles are defined by Vaudeney:

- *CreateTag()*: It generates a tag with unique identifiers and updates the database.
- *DrawTag()*: It selects or draws a set of fresh tags (which are to be sent to the adversary). It also identifies whether the drawn set of tags is legitimate.
- *Free()*: It releases a virtual tag (vtag), which then becomes unreachable by the adversary.
- *Launch()*: It launches a new protocol session.
- *SendReader()*: It sends a message to the reader during a protocol run.
- *Result()*: On the completion of a protocol run, it generates the output 0 or 1.
- *Corrupt()*: It returns tag status based on which tags are kept or destroyed.

Adversary: it is an algorithm that interacts with tags, has a public key, and requires oracles to carry out its operations. Vaudeney classified adversaries into the following categories:

- *Strong adversary*: It has access to all the oracles without any limitations.
- *Weak adversary*: It cannot corrupt tags, and it can only listen to the ongoing communications and interact with the oracle.
- *Forward adversary*: It becomes active once a tag is corrupted after which protocol interactions are not allowed.
- *Destructive adversary*: It cannot interact with a corrupted tag; that is, tags are destroyed after being corrupted.

Vaudeney stated that: $Strong^{Adv} \supseteq Destructive^{Adv} \supseteq Forward^{Adv} \supseteq Weak^{Adv}$

Security (Vaudeney's definition): If an adversary is able to identify a legitimate tag, then it wins the game. Thus, for a protocol to be secure, the probability of an adversary's winning should be negligible.

Privacy (Vaudeney's definition): During the game, the adversary has a set of drawn tags among which it has to identify a legitimate tag by the result 0 or 1. If the output is 1, then adversary wins; therefore, for a scheme to ensure privacy, all such adversaries should be trivial.

Blinder: A blinder B for any adversary A is a probabilistic polynomial time (PPT) algorithm that observes all the messages sent and received by A. It can also simulate *Launch()*, *SendReader()*, *SendTag()*, and *Result()* oracles. However, B is unable to access the reader's data.

An adversary is said to be blinded (denoted as A^B) if it does not use *Launch()*, *SendReader()*, *SendTag()*, and *Result()* oracles. Also, the adversary A^B is said to be trivial if after it is blinded, it produces the same output as before; that is, $|\Pr(A_{wins}) - \Pr(A^B_{wins})| \le \epsilon$, where ϵ is a negligible function.

The game between the adversary A and the challenger C is divided into two phases (as shown in Figure 7.1):

- *Attack phase*: During this phase, the adversary interacts with the system and queries with the oracle to gather information.
- *Analysis phase*: During this phase, the adversary is given a table of tags, among which it has to identify the legitimate and fake tags. A generates results as 0 or 1. If the output is 1, A wins.

Vaudeney's model [105] discusses various aspects of security and includes most of the correctness and privacy aspects. However, it has some weaknesses. Vaudeney stated that wide-strong privacy is not possible to achieve, which was later corrected by Ouafi and Vaudenay [106]. Vaudeney stated that there exists an entity called blinder, which answers the Result() oracle. Here, the adversary knows the selected tag, whereas the blinder doesn't. The blinder now knows the random coin flip.

7.1.2 Canard et al.'s Model

Canard et al. [107,108] proposed a security model, which is derived from Vaudeney's model. Canard et al.'s model uses strong, destructive, and weak adversaries. In addition, the use of oracles is also similar to those used by

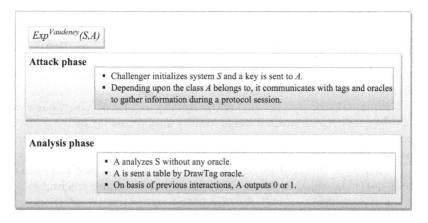

FIGURE 7.1   Vaudeney's privacy experiment.

Vaudeney's model. However, unlike the notion of security and privacy in Vaudeney's model, Canard et al. also ensure properties such as soundness, correctness, and untraceability.

Some of the new properties introduced by Canard et al. are as follows:

- Nonobvious link: a link between two virtual tags (v_1, v_2) is said to be nonobvious if
  i. Both tags refer to the same tag ID.
  ii. A "dummy" adversary that can call *DrawTag*, *CreateTag*, *Free*, *Corrupt*, and oracle is unable to generate this link with probability $\geq 1/2$.
- A nonobvious link can be categorized into the following classes:
  i. Standard: if both virtual tags are not corrupted by the adversary.
  ii. Past: if v_1 is uncorrupted.
  iii. Future: if v_2 is uncorrupted.
  iv. We have *Future(Strongest Privacy)* $\supseteq$ *Past* $\supseteq$ *Standard*.

Canard et al. [107] have dummy adversary A_D instead of a blinder B. Vaudeney has a blinder B, which is an entity on its own separate from the adversary A^B, whereas "dummy" adversary is a single entity having knowledge of all random choices that can be made.

Untraceability experiment: Untraceability can be proved in a system if all the adversaries are unable to generate a nonobvious link with probability greater than the dummy adversary (probability $\geq 1/2$). It can be stated as:

$$\text{Exp}^{\text{UNT}}(S,A) = \left| \Pr(A_{\text{wins}}) - \Pr(A_{D\text{wins}}) \right| \leq \in$$

Canard et al.'s model has most certainly overcome some of the demerits of Vaudeney's by using A_D and nonobvious links, but it still cannot be used with every scheme due to the use of nontrivial links and the limited scope within which untraceability is defined (as shown in Figure 7.2).

Correctness: A scheme S is correct for any class of adversary A, if any $A \in \{$Strong, Destructive, Weak$\}$ is a PPT algorithm where Success^{CORRECT} (S, A) is negligible (as shown in Figure 7.3). It can be stated as:

$$\text{Success}^{\text{CORRECT}}(S,A) = \Pr \left| \text{Exp}^{\text{CORRECT}}(S,A) = 0 \right|$$

> $Exp^{UNT}(S,A)$
>
> - Challenger initializes system S and a key is sent to A.
> - Depending upon the class A belongs to, it communicates with tags and oracles to gather information during a protocol session.
> - A returns a link (v_1, v_2)

FIGURE 7.2   Canard et al.'s untraceability experiment.

> $Exp^{CORRECT}(S,A)$
>
> - Challenger initializes system S and a key is sent to A.
> - Depending upon the class A belongs to, it communicates with tags and oracles to gather information during a protocol session.
> - A chooses an uncorrupted tag.
> - A launches the attack phase. The experiment returns a bit 'b' as generated by the oracle.

FIGURE 7.3   Canard et al.'s correctness experiment.

7.1.3 Universal Composability Framework

This framework [109,110] provides a very general methodology for executing and verifying protocol security, since this framework ensures security under composition operation; here, it is also referred to as universal composition. The protocols that satisfy the notions of this framework are said to be universally composable (UC).

Universal composition: The universal composition theorem states that running protocol π_ρ, with no access to F (where F is an ideal functionality), has essentially the same effect as running the original F-hybrid protocol π. More precisely, it guarantees that for any adversary A, there exists an adversary A_F such that no environment machine can tell with non-negligible probability whether it is interacting with A and parties running π_ρ, or with AF and parties running π. In particular, if π UC-realizes some ideal functionality G, then π_ρ can also do the same.

The UC framework is able to model the real world, the ideal world, and an emulation (that models a real-world protocol into an ideal world). It creates an interactive environment that captures the parameter values of the algorithm at the current time. The UC framework has the following components:

- In this framework, the adversary *A* interacts with environment *Z*, which comprises legitimate entities, each of which is PPT algorithms. *Z* generates the initial input parameters and obtains the output parameters by performing random queries with the adversary.

For an algorithm run for each adversary (PPT), there is a real-world simulation of the algorithm, which can be modeled into an ideal-world scenario in the presence of a simulated adversary *A* in such a way that *Z* cannot be able to distinguish if it is interacting with the adversary and algorithm instance in the ideal world or the real world.

The adversary in the UC security model interacts with both real-world and ideal-world environments. The adversary can eavesdrop into all the communications and can schedule the activation sequence of the entities. Every session has an identifier "sid," which is shared by all the entities involved in the communication. During the session initiation, *Z* becomes active first and then activates the adversary. If *Z* stops, the whole simulation is aborted. *Z* may also allow more than one protocol session to run simultaneously.

In the ideal scenario, the UC framework implements an anonymous authentication and anonymous key exchange functionality. To ensure anonymity in all the channels, anonymous wireless communications are involved. Figures 7.4 and 7.5 give a brief description of authentication and key exchange functionalities.

There are many ways for combining more than one protocol together, for example, sequential or concurrent executions. The protocol executions can be run by one or more than one entity and have one or more inputs and outputs. All of the scenarios can be considered as special cases of UC framework by varying uses of synchronization, sub-session, and inputs.

7.1.4 Juels–Weis Challenge–Response Model

The Juels–Weis [111] model is also a game-based framework where the system comprises one RFID reader R and n RFID tags. Each tag has its own key and pseudo-key, which is reset after every session. This model uses the following messages:

- To assign a new secret key to the tag, the *SetKey* message is used. When the tag receives *SetKey* message, it discards the old key value and a new arbitrary secret is allotted to the tag.
- The *TagInit* message initializes the session key to a new value, discards the current session details, and issues a new session key.
- The *ReaderInit* message is used by *R* to initialize a new session.

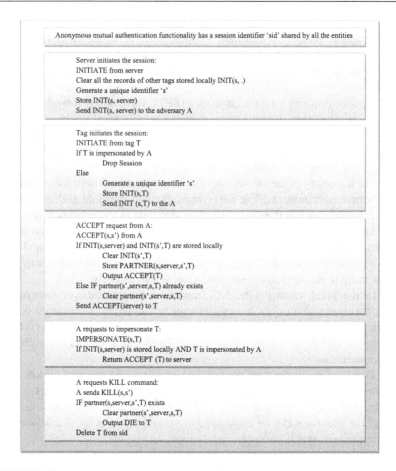

FIGURE 7.4 Anonymous mutual entity authentication with KILL command.

A is assumed to be an adversary which has the capability to generate any of the above messages. Once the tag is active, it is entitled to respond to any number of challenge–response messages, that is, (c_i, r_i) messages, which is based on the information from parameters such as previous sessions and challenge–response messages. The tag stores a log with the details of previous sessions and challenge–response message pairs.

Whenever the reader *R* receives the message (sid, r_i), it first evaluates a certain function based on its current status and all open as well as closed sessions. Based on this function, *R* outputs "accept" or "reject." *A* can corrupt any existing tag and issue any new tag as it is able to send any number of *SetKey* messages. Thus, if a tag receives *SetKey* message from *A*, it is said to be corrupted.

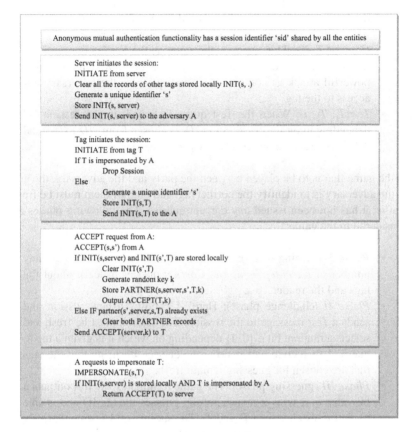

Anonymous mutual authentication functionality has a session identifier 'sid' shared by all the entities

Server initiates the session:
INITIATE from server
Clear all the records of other tags stored locally INIT(s, .)
Generate a unique identifier 's'
Store INIT(s, server)
Send INIT(s, server) to the adversary A

Tag initiates the session:
INITIATE from tag T
If T is impersonated by A
    Drop Session
Else
    Generate a unique identifier 's'
    Store INIT(s,T)
    Send INIT(s,T) to the A

ACCEPT request from A:
ACCEPT(s,s') from A
If INIT(s,server) and INIT(s',T) are stored locally
    Clear INIT(s',T)
    Generate random key k
    Store PARTNER(s,server,s',T,k)
    Output ACCEPT(T,k)
Else IF partner(s',server,s,T) already exists
    Clear both PARTNER records
Send ACCEPT(server,k) to T

A requests to impersonate T:
IMPERSONATE(s,T)
If INIT(s,server) is stored locally AND T is impersonated by A
    Return ACCEPT(T) to server

FIGURE 7.5   Anonymous key exchange.

In this approach, it is assumed that the adversary *A* listens and controls all the ongoing communications between the reader and the tags. Here, one reader *R* and a tags (*T* ∈ Tags), at the end of each session if one protocol party finds the other to be legitimate it outputs *Accept*.

A has the ability to issue the following queries:

- *Execute(R, T, i)*: Here, *A* eavesdrops on an actual execution of the protocol between the reader (*R*) and the tag (*T*) in session *i*. It is a passive attack.
- *Send(P_1, P_2, i, m)*: This query is the generalization of the challenge–response technique defined in Juels–Weis model along with the *TagInit* and *ReaderInit* messages. Here, the adversary can

impersonate a party P_1, which could be the reader or a tag during session "i," and sends m messages to some other party P_2.

- *Corrupt(T, K)*: This query is the same as the *SetKey* query. Here, the adversary changes the secret key of the tag to K. This is a more powerful attack as compared to Send query as the adversary has access to the tag.
- *Test(T_1, T_2, i)*: When the Test query is issued, in the session "i" depending upon $b \in \{0, 1\}$, an id ID_b is chosen from $\{ID_1, ID_2\}$ and A has to guess the bit "b" correctly to succeed.

In the game that is to be played between the party and the adversary, the goal of the adversary is to identify the correct tag, and both of them must be fresh; that is, it has not been issued any corrupt queries. The following phases are considered in the game:

- *Phase I* (learning phase): During this phase, A is able to send any number of *Execute*, *Send*, and *Corrupt* queries to learn about the tags and the reader.
- *Phase II* (challenge phase): Here, A chooses a new session and sends a *Test* message to the session. The session must be fresh and selects a random bit $\in \{0, 1\}$ depending on which it is given a tag to guess. A then continues making the queries, ensuring that the tags that are chosen for guessing remain fresh.
- *Phase III* (guessing phase): The game terminates when A outputs a bit value b; A wins if it successfully guesses the tag ID and is able to distinguish between T_0 and T_1. The success of A is quantitatively represented as an advantage of A and denoted as:

$$\text{Adv}_A^{\text{UNT}}(k) = \left| [A \text{ wins}] \Pr - \Pr [\text{random coin flip}] \right|$$

$$= \left| \Pr[b' = b] - \frac{1}{2} \right|$$

7.2 ISSUES WITH SECURITY MODELS

The real spreading of IoT services requires customized security and privacy levels to be guaranteed. The broad overview provided with this literature survey arises many open issues and sheds some light on research directions in the IoT security field. However, the domain still lacks a unified vision regarding the insurance of security and privacy requirements in a heterogeneous environment, involving different technologies and communication standards.

Suitable solutions need to be designed and deployed, which are independent from the exploited platform and able to guarantee confidentiality, access control, and privacy for users and things; trustworthiness among devices and users; and compliance with defined security and privacy policies.

In order to conduct holistic IoT trust management, we explored the trust properties that impact trust relationships, classified them into five categories, and indicated that holistic trust management should concern part or all of them in different contexts and for different purposes. Based on a general IoT system model, we presented some objectives for holistic IoT trust management and indicated their supporting IoT layers by emphasizing vertical trust management, which is crucial for achieving trust and security in IoT.

We have also discussed the security aspects of IoT with RFID as its enabling technology. Various security protocols have been proposed in this domain, and there are many methodologies that aid in providing a formal proof of correctness. Vaudeney's and Canard et al.'s models try to define privacy as indistinguishability between real-time and ideal-world scenarios. Vaudeney's model has offered more categories of the adversary, while other models consider maintaining privacy during authentication sessions. Vaudeney's model takes into account the privacy of all ongoing sessions. Considering real-time and ideal-world scenarios, the UC model can also analyze more than one protocol at a time, where the environment is also interactive. However, it may sometimes be a shortcoming as some insecure protocols may remain undetected. Thus, Vaudeney's model must be used if we want to ensure strong privacy.

Considering Juels–Weis model, privacy is ensured by the fact that the adversary is unable to distinguish between two tags. Canard et al.'s model, which is built up on Vaudeney's model, provides the same oracles, and they can easily implement the privacy notions of Juels–Weis model, which fall under its future-privacy notion. Thus, if we consider the indistinguishability of tags, Canard et al.'s model can ensure better privacy as compared to Juels–Weis model. Another issue with Juels–Weis model is that if we have to test a system having only one tag, the model fails as it requires at least two tags to be in a system for the challenge phase.

Some models are designed to be applied to certain categories of protocols, and they might not be able to model some other categories. For example, Vaudeney's and Canard et al.'s models can analyze authentication protocols well, and they can also be adapted to other models. However, the Juels–Weis model and the UC model allow all kinds of protocols to be modeled easily and are not restrictive in nature.

Both Vaudeney's and Canard et al.'s models provide most categories of adversaries as compared to others. Therefore, in order to model a weak adversary that cannot corrupt a tag under any circumstances (the possibility of this happening is very low), this model is feasible. Other models such as UC and Juels–Weis assume that a tag can always be corrupted. However, in Juels–Weis challenge phase, the adversary is unable to corrupt any of the tags. Vaudeney's model also provides an adversary with the authority to corrupt tags at the end of the protocol run under the forward secrecy notion. However, the Juels–Weis model has some constraints. It can also allow the adversary to corrupt a tag only to back-trace previous sessions that come under their notion of forward privacy.

Both Vaudeney's and Canard et al.'s models define a strong adversary, which has no restrictions on any kind of tag corruption. Initially, Vaudeney's model has failed to achieve the highest level of privacy, which can be reached in all other models. However, Ouafi et al. have extended Vaudeney's model to incorporate the notion of strong privacy.

UC allows more parallel and sequential runs of protocols and is able to incorporate all the tags present in the system at a time. However, Juels–Weis model specifies no tag at the initial stage, and the adversary is allowed to study $(n - 1)$ tags at most.

An Internet of Things (IoT)-Based Security Approach Ensuring Robust Location Privacy for the Healthcare Environment

8

Recently, various applications of IoT are being developed for the healthcare domain. In this chapter, our contribution is to design a lightweight security approach for the Internet of Things (IoT) devices in healthcare. The main goal is to make patients' lives easier and comfortable and to provide them more effective treatment. In addition, we also intend to address the issues related to location privacy and security due to the deployment of IoT devices. We have proposed a very simple mutual authentication protocol, which provides strong location privacy using hash function, pseudo-random number generator (PRNG), and bitwise operation. The security strength of our protocol is verified through a formal proof model for location privacy.

8.1 PROBLEM DEFINITION

In order to ensure privacy in the healthcare domain, the identity of tags must remain a secret to all unauthorized parties. Tag's identity is disclosed to authorized readers only. Every reader should authenticate itself to the tag before any communication. The location privacy can be protected if the tag identifier (ID) is protected. The ID is a fixed integer value set during manufacturing that uniquely identifies a tag. This ID is permanent until the tag dies or is destroyed. Sometimes tag ID leaves certain traces analyzing which the adversary can find out the tag ID. The disclosure of tag ID poses a serious threat to devices in a network. Therefore, we have designed a very simple mutual authentication algorithm, which aims to secure the tag ID.

The proposed solution is a challenge response-based scheme. In this case, to ensure location privacy, we have made sure that our solution provides indistinguishability and forward secrecy. First, we ensure that the adversary is not able to identify which tag is communicating with the reader, that is, indistinguishability. Second, we ensure that in case the current secret parameters are revealed, it does not compromise the security of earlier parameters.

8.2 ABSTRACT OVERVIEW

IoT devices are needed to be protected from a wide range of threats, which include malware infections, disruption of services, and information theft. The attacker could easily gain control of the devices that are a part of smart home, automobiles, personal fitness, disease-monitoring gadgets, etc. An attacker can simply hack the software in a person's smart watch or insulin pump to track their location, or they might gain access to the information systems present in the automobiles and use them to carry malicious activities. Improving the efficiency of medical infrastructure is a very challenging task at present. The developments in IoT such as fully automated device identification and tracking and monitoring of patients in real time are helping to deliver quality healthcare services. These services also help in reducing the cost and at the same time reducing the resource requirements, which is a big issue in most of the hospitals.

One of the major concerns for such networks is the security and privacy of the devices. In the healthcare environment, data privacy is a major concern as the data on these devices is mostly patients' personal information. Thus, it is crucial to mitigate the security risks and to develop strong and secure networks keeping in mind the cost barrier.

IoT-based healthcare systems are networks containing devices that are interconnected and perform sharing and processing of a significant amount of data. Securing these networks requires cryptographic techniques, which can be adapted as per the architecture and computing capabilities of IoT devices. Various such protocols have been proposed in the past few years. One of the major security concerns is that one single device cannot be considered alone because even if a single device in the network is vulnerable to outsider threats, it will provide a gateway for unauthorized access to the whole network. To remedy this issue, we need to ensure that all the devices in the network are secure against tracking and disclosure attacks so that the device identification, location, and communication data remains secure.

In this chapter, we propose a simple lightweight mutual authentication protocol for healthcare systems with low computation and storage requirements. Our protocol uses a minimal number of messages between the reader and the tag for authentication purposes. We deploy one-way hash, random number generator, and bitwise shift operation for our protocol, which ensures mutual authentication between the tag and the reader. Location privacy is one of the primary requirements for any device. Our protocol ensures strong security by providing both indistinguishability and forward secrecy. We have deployed a formal model to give a demonstration of security provided by the proposed protocol to ensure security, privacy, and correctness.

8.3 DETAILED PROTOCOL DESCRIPTION

In order to ensure the security of an IoT system, we need to ensure the basic security requirements at the radio frequency identification (RFID) level. These are considered to be basic criteria that must be fulfilled by the protocol to protect security attacks on a system.

8.3.1 Location Privacy-Based Mutual Authentication Protocol

Location privacy is one of the most prominent threats to IoT devices. In case of RFID tags, the message exchange between the server and the tag is done via radio frequency, thus making it easier for attackers to extract the information during communication. If the messages are not properly secured or encrypted, it might lead to the infringement in the devices' location privacy. Each RFID tag has information for computation purposes; thus, security issues are there automatically. In healthcare sector, the problem of location tracking is of utmost concern as it may disclose a person's private and sensitive information. By being able to trace a tag's location, the adversary might also detect a patient's movements and activities, which poses a very serious threat.

For any protocol to ensure strong location privacy, two conditions need to be satisfied [111]:

- *Indistinguishability*: The parameters sent by various tags cannot be distinguished by any adversary.
- *Forward secrecy*: It ensures that even if the adversary obtains secret parameters of a tag, they are not able to track its location with the help of messages from previous sessions.

8.3.2 Authentication of IoT Devices

Before initiating a communication with a new device, users must ensure that it is legitimate (either from the hospital or from the patient). Therefore, the requesting party should initialize an authentication session before any information transfer. The hosting party, if legitimate, will then correctly authenticate itself using random challenge parameters. To ensure security, the ID of the devices must always be protected. Whenever a party is required to send its ID value, proper encryption must be done. For such requirements, various lightweight cryptographic mechanisms are used.

8.4 IoT AUTHENTICATION SCHEME ENSURING LOCATION PRIVACY

In the previous sections, we have briefly discussed the current scenario of IoT-based healthcare and security requirements it entails. Now, we will discuss the idea of a simple authentication protocol to ensure location privacy for devices in healthcare (Figure 8.1).

Setup: Here, tuple (*ID*, *k*) of the devices are stored by the reader. The devices and the reader also have PRNGs for generating fresh nonce values for authentication in every session. The messages in transit are encrypted using one-way hash (*h*()). The attacker analyzes the traffic for secret information,

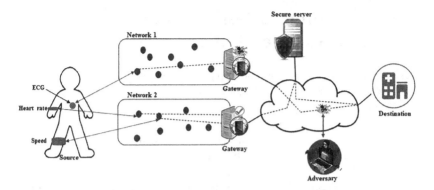

FIGURE 8.1   Secure communication among IoT sensors and server.

which is sent via radio frequency, and records the communication between the server and the tag. However, it is not possible for the attacker to decrypt the hash values; therefore, each message has a new random variable value. For each tag, these values will be different; thus, the intercepted information is of no use to the attacker (Figure 8.2).

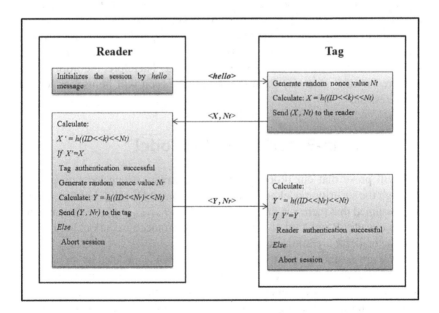

FIGURE 8.2   Flowchart of our proposed solution.

8.4.1 Concept of a Basic Location Privacy-Based Authentication Scheme

i. The reader sends a session initialization message to the tag.
ii. The tag generates a random nonce value N_t, calculates $X = h((ID \ll k) \ll N_t)$, and then sends (X, N_t) to the reader.
iii. On receiving tag's response, the reader now calculates $X' = h((ID \ll k) \ll N_t)$ and compares it with X. If $X' = X$, then the reader generates another random nonce value N_r, calculates $Y = h((ID \ll N_r) \ll N_t)$, and sends (Y, N_r) to the tag.
iv. On receiving the response from the reader, if the tag is able to calculate the value of Y and it matches the value received, then the session will be successful. If the computed value Y does not match the received value, then the session is immediately dropped.

8.5 SECURITY ANALYSIS

There may be the case where the adversary tries to impersonate the server and sends a captured value of N_r to initiate the session. Since the nonce values are fresh for every session, the tag computes a new N_t and calculates X, even if the attacker knows both N_r and N_t. However, ID is still unknown, and for the next session, both nonce values will be different. Therefore, the information gathered by the adversary from the previous session will be useless for the fresh session.

8.5.1 Game-Based Security Model

We will prove that our proposed approach exhibits strong location privacy through the security model proposed by Chawla and Ha [112]. For this model, it is assumed that there is a finite set of tags $T = \{t_1, t_2,..., t_n\}$. The system comprises a reader R and a set of tags. Prior to the protocol execution, both tags T and the reader R are initialized.

We are assuming an active adversary A, which is modeled by an oracle O which has control over all the communications going on within the network. The oracle receives the query message and generates appropriate outputs. Table 8.1 shows the set of query messages involved during this model.

TABLE 8.1   List of Query Messages

QUERY	DESCRIPTION
Query(R, m_1)	Invokes R and outputs a message m_1
Reply(R/T_i, m_1, m_2)	Invokes tag or reader on receiving m_1 and responds with message m_2
Auth(p_1, p_2)	For this query, the oracle models perform the calculation of parameter p_2 after successfully identifying p_1
Execute(R, T_i)	The oracle models A to eavesdrop during an ongoing session to extract all the messages. Execute(R, T_i) = {Query(R, m_1). Reply(R/T_i, m_1, m_2). Auth(p_1, p_2)}
Reveal(T_i)	Invokes all the secret parameters of a tag
Test(T_i)	To check whether the adversary is successful or not.

We assume the adversary to be active, and therefore, it is able to execute any of the above-mentioned queries on all the tags any number of times. However, *test*() can be executed only once on a single tag. We will execute an attack game on our protocol to prove its security:

Phase I: Initialization
1. Set each tag $t_i \in T$ with secret key k_i.
2. Save $k_i \ \forall \ t_i \in T$ in the reader's database.

Phase II: Learning
1. A executes oracle $O \ \forall \ t_i \in T$ present in the system except for any one tag t_c, which is saved for the challenge phase.
2. A calls Query(R, m_1), Reply(R/T_i, m_1, m_2), Auth(p_1, p_2), and Reveal(T_i) on all the tags any number of times.

Phase III: Challenge
1. A selects t_c from T for the challenge phase.
2. A executes oracle O for tag t_c.
3. A calls Query(R, m_1), Reply(R/T_i,m_1,m_2), and Auth(p_1,p_2), but not Reveal(T_i), on any number of times.
4. A executes oracle O for Test(t_c).
   - Oracle O tosses a fair coin $b \in \{0, 1\}$; let $b \leftarrow \{0, 1\}$.
   - If $b = 1$, A correctly identifies t_c.
   - If $b = 0$, A calculates a random value.
5. A executes oracles for $n - 1$ tags $t_i \in T$ except t_c.
6. A outputs a guess bit b'.
7. A wins the attack game if $b = b'$.

8.5.2 Strong Location Privacy

Definition 1: *An authentication algorithm is said to exhibit strong location privacy if it ensures both indistinguishability and forward secrecy ∀ adversaries (passive or active) with queries e, r, a, x (where e =* Execute(); *r =* Reply(); *a =* Auth(); *and x is the secret parameter*).

a. *Proof of indistinguishability*: For A to successfully determine the location of t_c, A must determine (ID_c, k_c) from all the guessed values generated on the basis of the knowledge A obtained during the learning phase. In order to distinguish between the tag values, A must be able to successfully determine that some values $(ID'_c, k'_c, N'_t, N'_r)$ are not corresponding to t_c. Moreover, A should be able to know that $(ID_c \neq ID'_c, k_c \neq k'_c, N_t \neq N'_t, N_t \neq N'_r)$. Suppose A executes at most α Execute() queries, β Reply() queries, and γ Auth() queries for t_c during the challenge phase. For each session, there is a generation of fresh random nonce values and key values having 2^q space (assuming each parameter is q-bits in size). Thus, the condition for A to distinguish the values correctly has the probability $p = \dfrac{(\alpha + \beta + \gamma)}{2^q}$, which is negligible; thus, our protocol exhibits indistinguishability.

b. *Proof of forward secrecy*: During the challenge phase, suppose A makes α Execute() queries ∀ $(n-1)$ tags $t_i \in T$. Here, the advantage of A guessing the ID values for all the tags based on the knowledge from the learning phase is denoted as $\text{Adv}_A^g(t_i)$, which is as follows:

$$\text{Adv}_A^g(t_i) = \left(\frac{1}{2^k}\right)^e$$

A has a random value or t_c's real parameters on the basis of (ID_c, k_c, N_t, N_r). Therefore, the aim of A is to extract the ID_c value from t_c using the execute queries for eavesdropping and knowledge from the other tags and previous sessions. The advantage of A correctly guessing the fair coin value b denoted as $\text{Adv}_A^{fs}(t_c)$ is calculated as follows:

$$\text{Adv}_A^{fs}(t_c) \leq \text{Adv}_A^g(t_1) + \text{Adv}_A^g(t_2) + \cdots + \text{Adv}_A^g(t_{n-1})$$

$$\leq (n-1)\operatorname{Adv}_A^g (t_1)$$

$$\leq (n-1)\left(\frac{1}{2^k}\right)^e \leq (n-1)\left(\frac{1}{2^{ke}}\right)$$

which is negligible; thus, our protocol ensures forward secrecy.

Since it has been proved that the above protocol ensures both indistinguishability and forward secrecy, it exhibits strong location privacy.

8.6 PERFORMANCE ANALYSIS

It is known that IoT-based devices have limited resources, which enables us to take into account performance issues in real-world applications. Authentication protocols must consider the cost of all the computations and message communications. The use of random numbers adds to our protocol's security and reduces the storage cost.

Our protocol requires a tag to store three parameters: the device ID, the session key k, and the random number N_r. Each is 96-bits in size. The total storage cost of our protocol is 286 (3×96) bits.

The tag uses a one-way hash function with time complexity $O(1)$ and random number generator also having complexity $O(1)$. The complexity of the circular left shift is $w \times O(n)$, where w is the hamming weight of the parameter with respect to which the value is to be rotated. Circular shift operation also requires 1-bit auxiliary space.

8.7 CONCLUSION

The increasing cost of healthcare is a big issue these days as everyone cannot afford the resources. The main aim behind the integration of IoT with healthcare is to provide healthcare facilities to anyone at any place. Although the devices may be large or may be as small as rice grain installed inside a human being or an animal, it is important to secure these devices as they deal with human-related data and vitals. In this chapter, we have presented a very simple yet strong location privacy authentication solution for such devices, which can be implemented at very low cost without affecting the device mobility as it does not require any higher-layer resources. Our protocol is suitable for both passive and active tags. Further, our work can be extended to make the proposed solution suitable for other business models, taking into account their functional and nonfunctional requirements. At the physical level, we can work on variants of our protocol to be based on some hard-mathematical problems.

References

1. J. Buckley (ed.) *The Internet of Things: From RFID to the Next-Generation Pervasive Networked Systems.* Auerbach Publications, New York, 2006.
2. L. Atzori, A. Iera, G. Morabito, "The Internet of Things: a survey," *Comput. Netw.* vol. 54, pp. 2787–2805, 2010.
3. M. Zorzi, A. Gluhak, S. Lange, A. Bassi, "From today's Intranet of Things to a future Internet of Things: a wireless- and mobility-related view," *IEEE Wirel. Commun.* vol. 17, pp. 43–51, 2010.
4. H.S. Ning, Z.O. Wang, "Future Internet of Things architecture: Like mankind neural system or social organization framework?" *IEEE Commun. Lett.* vol. 15, pp. 461–463, 2011.
5. K.E. Psannis, S. Xinogalos, A. Sifaleras, "Convergence of Internet of Things and mobile cloud computing," *Syst. Sci. Control Eng. Open Access J.* vol. 2, no. 1, pp. 476–483, 2014.
6. R. Roman, P. Najera, J. Lopez "Securing the Internet of Things," *Computer* vol. 44, no. 9, 51–58, 2011.
7. J. Gubbi, R. Buyya, S. Marusic, M. Palaniswami, "Internet of Things (IoT): a vision, architectural elements, and future directions," *Futur. Gener. Comput. Syst.* vol. 29, no. 7, pp. 1645–1660, 2013.
8. K. Ashton, "That 'Internet of Things' thing," *RFiD J.* 2009. www.itrco.jp/libraries/RFIDjournalThat%20Internet%20of%20Things%20Thing.pdf, Last accessed January 2019.
9. Near Field Communications History "Timeline of RFID technology," 2016. Available at: www.nfcnearfieldcommunication.org/timeline.html, Last accessed January 2019.
10. Postscapes, "History of Internet of Things," 2016. Available at: http://postscapes.com/internet-of-things-history, Last accessed January 2019.
11. E. Welbourne, L. Battle, G. Cole et al. "Building the Internet of Things Using RFID: the RFID ecosystem experience," IEEE Computing Society, 2009. Available at: http://homes.cs.washington.edu/~magda/papers/welbourne-ieeeic09.pdf, Last accessed January 2019.
12. G. Abramovich, "15 facts about Internet Of Things," April 2015, Available at: www.cmo.com/articles/2015/4/13/mind-blowing-stats-Internet-of-things-iot.html, Last accessed January 2019.
13. J. Hanson, "The 10 challenges of securing IoT communications," May 2015, Available at: www.pubnub.com/blog/2015-05-04-10-challenges-securing-iot-communications-iot-security/, Last accessed January 2019.
14. Cyber Security Ventures, "IoT Security Report, Q3 2015." Available at: http://cybersecurityventures.com/internet-of-things-security-report-q3-2015/, Last accessed April 2016.

15. Sophos Labs, "Mobile and IoT attacks: SophosLabs 2019 Threat Report." Available at: https://nakedsecurity.sophos.com/2018/11/23/mobile-and-iot-attacks-sophoslabs-2019-threat-report/, Last accessed January 2019.

16. Forbes, "2017 Roundup of Internet Of Things forecasts." Available at: www.forbes.com/sites/louiscolumbus/2017/12/10/2017-roundup-of-internet-of-things-forecasts/#2bbbc6541480, Last accessed January 2019.

17. IT Pro, "IoT malware tripled in the first half of 2018." Available at: www.itpro.co.uk/malware/31945/iot-malware-tripled-in-the-first-half-of-2018, Last accessed January 2019.

18. J. Granjal, E. Monteiro, and J. Sá Silva, "Security for the Internet of Things: A survey of existing protocols and open research issues," *IEEE Commun. Surveys Tuts.*, vol. 17, no. 3, pp. 1294–1312, 3rd Quart., 2015.

19. IEEE Standard for Information Technology, Telecommunications and information exchange between systems-Local and metropolitan area networks-Specific requirements Part 15.4: Wireless Medium Access Control (MAC) and Physical Layer (PHY) Specifications for Low-Rate Wireless Personal Area Networks (WPANs) Amendment 3: Alternative Physical Layer Extension to support the Japanese 950 MHz bands, IEEE Std 802.15.4d-2009 (Amendment to IEEE Std 802.15.4-2006), (2009) 1–27 doi:10.1109/IEEESTD.2009.4840354.

20. S. Raza, S. Duquennoy, T. Voigt, "Securing communication in 6LoWPAN with compressed IPsec," in *International Conference on Distributed Computing in Sensor Systems (DCOSS)*, Barcelona, 1–8 2011. doi:10.1109/DCOSS.2011.5982177.

21. T. Clausen, U. Herberg, and M. Philipp, "A critical evaluation of the IPv6 routing protocol for low power and lossy networks (RPL)," in *Proceedings of the IEEE 7th International Conference on Wireless and Mobile Computing*, Istanbul, 2011, pp. 365–372.

22. Z. Shelby, K. Hartke, C. Bormann, and B. Frank, *Constrained Application Protocol (CoAP).draft-ietf-core-coap-18*. Internet Eng. Task Force (IETF), Fremont, CA, 2013.

23. W. Colitti, K. Steenhaut, N. De Caro, B. Buta, and V. Dobrota, "Evaluation of constrained application protocol for wireless sensor networks," in *Proceedings of the 18th IEEE Workshop on Local and Metropolitan Area Networks (LANMAN)*, Chapel Hill, NC, 2011, pp. 1–6.

24. T. Borgohain, U. Kumar, S. Sanyal, "Survey of operating systems for the IoT environment," arXiv preprint arXiv:1504.02517, 2015/4/13.

25. Arm Limited MBed OS, Available at: https://mbed.org/technology/os/ Last accessed February 2018.

26. E. Baccelli, O. Hahm, M. Günes, M. Wählisch, and T.C. Schmidt, "RIOT OS: Towards an OS for the Internet of Things," in *Proceedings of the IEEE Conference on Computer Communications Workshops (INFOCOM WKSPS)*, Turin, Italy, 2013, pp. 79–80.

27. A. Dunkels, B. Gronvall, and T. Voigt, "Contiki—A lightweight and flexible operating system for tiny networked sensors," in *Proceedings of the 29th Annual IEEE International Conference on Local Computer Networks*, Tampa, FL, 2004, pp. 455–462.

28. A. Eswaran, A. Rowe, and R. Rajkumar. "Nano-rk: An energy-aware resource-centric RTOS for sensor 5 networks," in *26th IEEE International Real-Time Systems Symposium (RTSS'05)*, Miami, FL, IEEE, 2005, p. 10.

29. P. Barnaghi, W. Wang, C. Henson, and K. Taylor, "Semantics for the Internet of Things: Early progress and back to the future," *Proc. IJSWIS*, vol. 8, no. 1, pp. 1–21, January 2012.

30. T. Kamiya and J. Schneider, *"Efficient XML Interchange (EXI) Format 1.0,"* World Wide Web Consortium, Cambridge, MA, 2011.

31. S. De, B. Christophe, and K. Moessner, "Semantic enablers for dynamic digital-physical object associations in a federated node architecture for the Internet of Things," *Ad Hoc Netw.*, vol. 18, pp. 102–120, 2014.

32. L. Yu and Y. Liu, "Using the linked data approach in a heterogeneous sensor web: Challenges, experiments and lessons learned," in *Proceedings of Sensor Web Enablement (SWE) Workshop*, Banff, Alberta, Canada, 2011.

33. F. Li, M. Voegler, M. Claessens, and S. Dustdar, "Efficient and scalable IoT service delivery on cloud," in *Proceedings of the IEEE 6th International Conference on CLOUD*, Dresden, 2013, pp. 740–747.

34. C. Wang, Z. Bi, and L.D. Xu, "IoT and cloud computing in automation of assembly modeling systems," *IEEE Trans. Ind. Informat.*, vol. 10, no. 2, pp. 1426–1434, May 2014.

35. B. Rochwerger et al., "The reservoir model and architecture for open federated cloud computing," *IBM J. Res. Develop.*, vol. 53, no. 4, pp. 535–545, July 2009.

36. G.C. Fox, S. Kamburugamuve, and R.D. Hartman, "Architecture and measured characteristics of a cloud based Internet of Things," in *Proceedings of the International Conference on CTS*, Denver, CO, 2012, pp. 6–12.

37. Gartner, "The Importance of 'Big Data': A Definition," June 21 2012, www.gartner.com/doc/2057415/ importance-big-data-definition, Last accessed June 2014.

38. C. Hsinchun, R.H.L. Chiang, and V.C. Storey, "Business intelligence and analytics: From big data to big impact," *MIS Quarterly*, vol. 36, no. 4, pp. 1165–1188, 2012.

39. H. Jiang, F. Shen, S. Chen, K.-C. Li, Y.-S. Jeong, "A secure and scalable storage system for aggregate data in IoT," *Future Generation Comp. Syst.* vol. 49, 133–141, 2015.

40. F. Bonomi, R. Milito, J. Zhu, and S. Addepalli, "Fog computing and its role in the Internet of Things," in *Proceedings of the 1st Edition of the MCC Workshop on Mobile Cloud Computing*, Helsinki, Finland, 2012, pp. 13–16.

41. M. Yannuzzi, R. Milito, R. Serral-Gracia, and D. Montero, "Key ingredients in an IoT recipe: Fog computing, cloud computing, and more fog computing", in *International Workshop on Computer Aided Modeling and Design of Communication Links and Networks (CAMAD)*, Athens, Greece, 2014, pp. 325–329.

42. W. Shi, J. Cao, Q. Zhang, Y. Li, and L. Xu, "Edge computing: Vision and challenges," *IEEE Internet Things J.*, vol. 3, no. 5, pp. 637–646, October 2016.

43. R. Mahmud and R. Buyya. (2016). "Fog computing: A taxonomy, survey and future directions." (Online). Available at: http://arxiv. org/abs/1611.05539.

44. N. Wu, M. Nystrom, T. Lin, and H. Yu, "Challenges to global RFID adoption," *Technovation*, vol. 26, pp. 1317–1323, 2006.

45. R. Ratasuk, A. Prasad, Z. Li, A. Ghosh, and M.A. Uusitalo. "Recent advancements in M2M communications in 4G networks and evolution towards 5G," in *Innovation in Clouds, Internet and Networks*, Paris, 2015, pp. 52–57.

46. B. Ray, "What is M2M? 2018 Update," Available at: www.link-labs.com/blog/what-is-m2m, Last accessed September 2019.

47. Business Wire, "Internet of Things (IoT) & Machine-to-Machine (M2M) industry almanac, 2019." Available at: www.businesswire.com/news/home/20190501005539/en/Internet-Things-IoT-Machine-to-Machine-M2M-Industry-Almanac, Last accessed September 2019.

48. A. Holst, "M2M (machine-to-machine) - statistics & facts." Available at: www.statista.com/topics/1843/m2m-machine-to-machine/, Last accessed September 2019.

49. K. Fischer and J. Gesner, "Security architecture elements for IoT enabled automation networks," in *2012 IEEE 17th Conference on Emerging Technologies Factory Automation (ETFA)*, Krakow, Poland, September 2012, pp. 1–8.

50. S. Pirbhulal, H. Zhang, M.E. E Alahi, H. Ghayvat, S.C. Mukhopadhyay, Y.T. Zhang, W. Wu, "A novel secure IoT-based smart home automation system using a wireless sensor network," *Sensors*, vol. 17, p. 69, 2016. doi:10.3390/s17010069.

51. Z. Pang, Q. Chen, J. Tian, "Ecosystem analysis in the design of open platform-based in-home healthcare terminals towards the Internet-of-Things," In *Proceedings of the 15th International Conference on Advanced Communication Technology (ICACT)*, PyeongChang, Korea, 27–30 January 2013.

52. Y. Zhang, B. Chen, and X. Lu, "Intelligent monitoring system on refrigerator trucks based on the Internet of Things," *Wireless Commun. Appl.*, vol. 72, pp. 201–206, 2012.

53. Z. Ji and A. Qi, "The application of Internet of Things (IoT) in emergency management system in China," in *Proceedings of the 2010 IEEE International Conference on Technologies for Homeland Security (HST)*, Waltham, MA, 2010, pp. 139–142.

54. EPIC, "Internet of Things (IoT)." Available at: https://epic.org/privacy/internet/iot/, Last accessed September 2019.

55. GAO, "INTERNET OF THINGS: Status and implications of an increasingly connected world." Available at: https://www.gao.gov/products/GAO-17-75, Last accessed September 2019.

56. T. Macaulay, *RIoT Control: Understanding and Managing Risks and the Internet of Things*. Elsevier (USA), 2016.

57. S. Gürses, B. Berendt, T. Santen, "Multilateral security requirements analysis for preserving privacy in ubiquitous environments," in *Proceedings of the UKDU Workshop*, Berlin, 2006, pp. 51–64.

58. X. Huang, R. Fu, B. Chen et al. "User interactive Internet of Things privacy preserved access control," in *2012 International Conference for Internet Technology and Secured Transactions*, IEEE, 2012, pp. 597–602.

59. J. Cao, B. Carminati, E. Ferrari et al. "Castle: Continuously anonymizing data streams," *IEEE Trans. Depend. Secure Comput.*, vol. 8, pp. 337–352, 2011.

60. A. Alcaide, E. Palomar, J. Montero-Castillo et al. "Anonymous authentication for privacy-preserving IoT target-driven applications," *Comput. Secur.*, vol. 37, pp. 111–123, 2013.
61. E.-O. Blass, K. Elkhiyaoui, R. Molva, "Tracker: Security and privacy for RFID-based supply chains," in *NDSS'11, 18th Annual Network and Distributed System Security Symposium*, San Diego, CA, 6–9 February 2011.
62. K. Elkhiyaoui, E.-O. Blass, R. Molva (2012). "CHECKER: On-site checking in RFID-based supply chains," in *Proceedings of the Fifth ACM Conference on Security and Privacy in Wireless and Mobile Networks*, ACM, Florida, pp. 173–184.
63. H. Sundmaeker, P. Guillemin, P. Friess et al. "Vision and challenges for realising the Internet of Things," *Cluster of European Research Projects on the Internet of Things, European Commision*, vol. 3, no. 3, pp. 34–36, 2010.
64. S. Hachem, T. Teixeira, V. Issarny, "Ontologies for the Internet of Things," in *Proceedings of the 8th Middleware Doctoral Symposium*, ACM, Lisbon, Portugal, 2011, p. 3.
65. J. Wan, H. Yan, H. Suo et al. (2011). "Advances in cyber-physical systems research," *TIIS* vol. 5, pp. 1891–1908.
66. Y. Wang, Q. Wen, "A privacy enhanced DNS scheme for the Internet of Things," in *IET International Conference on Communication Technology and Application (ICCTA 2011)*, *IET*, Beijing, 2011, pp. 699–702.
67. Tech Target, "IoT security trends to watch in 2019," Available at: https://internetofthingsagenda.techtarget.com/blog/IoT-Agenda/IoT-security-trends-to-watch-in-2019, Last accessed September, 2019.
68. E. Bertino, "Data Security and Privacy in the IoT," in *Proceedings of the Extending Database Technology (EDBT)*, Bordeaux, France, March 2016.
69. S. Sultana, D. Midi, and E. Bertino. "Kinesis: A security incident response and prevention system for wireless sensor networks," in *Proceedings of the 12th ACM Conference on Embedded Network Sensor Systems, SenSys '14*, Memphis, TN, November 3–6 2014, pp. 148–162.
70. K. Zhao and L. Ge, "A survey on the Internet of Things security," in *2013 9th International Conference on Computational Intelligence and Security (CIS)*, Lehsan, China, December 2013, pp. 663–667.
71. Y. Chen, J. Chou, and C. Lin, "A novel RFID authentication protocol based on elliptic curve cryptosystem," Cryptology ePrint Archive, Report 2011/381, 2011.
72. T. Yeh, Y. Wanga, T. Kuo, S. Wang, "Securing RFID systems conforming to EPC Class 1 Generation 2 standard," *Exp. Syst. Appl.* vol. 37, pp. 7678–7683, 2010.
73. J. Cho, S. Yeo, and S. Kim, "Securing against brute-force attack: A hash-based RFID mutual authentication protocol using a secret value," *Comput. Commun.*, vol. 34, no. 3, pp. 391–397, 2011.
74. M. Burmester, T. van Le, and B. de Medeiros. "Provably secure ubiquitous systems: Universally composable RFID authentication protocols," 2006. Referenced 2006 at http://eprint.iacr.org/2006/131.pdf.
75. Y. Liao and C. Hsiao, "A secure ECC-based RFID authentication scheme using hybrid protocols," in Y. Liao and C. Hsiao (eds.) *Advances in Intelligent Systems and Applications*, Springer-Verlag, Berlin, Germany, pp. 1–13, 2013.

76. J. Chou, "An efficient mutual authentication RFID scheme based on elliptic curve cryptography," *J. Supercomput.*, vol. 70, no. 1, pp. 75–94, 2014.

77. Y.-J. Tu and S. Piramuthu. "RFID distance bounding protocols," in *1st International EURASIP Workshop in RFID Technology*, Vienna, Austria, 2007.

78. M. Ohkubo, K. Suzuki, and S. Kinoshita, "Cryptographic approach to privacy-friendly tags," in *RFID Privacy Workshop*, New Jersey, 2003.

79. S.A. Weis, S.E. Sarma, R.L. Rivest, D.W. Engels, "Security & privacy aspects of low-cost radio frequency identification systems," Security in Pervasive Computing, LNCS no. 2802, 2004, pp. 201–212.

80. A. Henrici, P. Muller, "Hash-based enhancement of location privacy for radio-frequency identification devices using varying identifiers," in *International Workshop on Pervasive Computing and Communication Security PerSec*, Orlando, FL, 2004, pp. 149–153.

81. D. Molnar, D. Wagner, "Privacy and security in library RFID: Issues, practices, and architectures," in *Conference on Computer and Communications Security—ACM CCS*, Washington, DC, 2004, pp. 210–219. (ISBN: 1-58113-961-6).

82. P. Tuyls, L. Batina, "RFID-tags for anti-counterfeiting," In Topics in Cryptology (CT-RSA'06), LNCS 3860, 2006, pp. 115–131.

83. Y.K. Lee, L. Batina, and I. Verbauwhede, "EC-RAC (ECDLP based randomized access control): Provably secure RFID authentication protocol," in *IEEE International Conference on RFID*, Las Vegas, NV, 2008, pp. 97–104.

84. Y. Liao and C. Hsiao, "A secure ECC-based RFID authentication scheme integrated with ID-verifier transfer protocol," *Ad Hoc Networks*, 2013, doi:10.1016/j.adhoc.2013.02.004.

85. R. Peeters and J. Hermans, "Attack on Liao and Hsiao's secure ECC-based RFID authentication scheme integrated with ID-verifier transfer protocol," Cryptology ePrint Archive, Report 2013/399, 2013.

86. Z. Tan, "A user anonymity preserving three-factor authentication scheme for telecare medicine information systems," *J. Med. Syst.* vol. 38, no. 3, pp. 1–9, 2014.

87. H. Arshad and M. Nikooghadam, "Three-factor anonymous authentication and key agreement scheme for telecare medicine information systems," *J. Med. Syst.* vol. 38, no. 12, pp. 1–12, 2014.

88. Y. Lu, L. Li, H. Peng, Y. Yang, "An enhanced biometric-based authentication scheme for telecare medicine information systems using elliptic curve cryptosystem," *J. Med. Syst.* vol. 39, no. 3, p. 32, 2015. doi:10.1007/s10916-015-0221-7.

89. J. Delvaux, D. Gu, I. Verbauwhede, M. Hiller, and M.-D. Yu, "Efficient fuzzy extraction of PUF-induced secrets: Theory and applications," in *Proceedings of the 18th International Conference on Cryptographic Hardware and Embedded Systems (CHES)*, vol. 9813, Santa Barbara, CA, August 2016, pp. 412–431.

90. M. Akgun and M. U. Caglayan, "Providing destructive privacy and scalability in RFID systems using PUFs," *Ad Hoc Netw.*, vol. 32, pp. 32–42, September 2015.

91. D. Sadhya and S.K. Singh, "Biometric Template Security and Biometric Encryption Using Fuzzy Frameworks," in *Encyclopedia of Information Science and Technology*, Third Edition. IGI Global (USA), pp. 512–524, 2015.

92. D. Sadhya and S.K. Singh, "Providing robust security measures to Bloom filter based biometric template protection schemes," *Comput. Secur.*, vol. 67, pp. 59–72, 2017.

93. E. Aysu, D. Gulcan, P. Moriyama, Schaumont, and M. Yung, "End-to-end design of a PUF-based privacy preserving authentication protocol," in *Proceedings of the 17th International Conference on Cryptographic Hardware and Embedded Systems (CHES)*, vol. 9293, Saint-Malo, France, September 2015, pp. 556–576.

94. N.J. Hopper and M. Blum, "Secure human identification protocols, Advances in Cryptology – ASYACRYPT'2001," Lecture Notes in Computer Science, vol. 2248, Springer, 2001, pp. 52–66.

95. M.L. Blum, M.J. Furst, R.J. Kearns, "Lipton, crypto-graphic primitives based on hard learning problems, Advances in Cryptology – CRYPTO'93," Lecture Notes in Computer Science, Springer, 1993, pp. 278–291.

96. A. Aysu, Y. Wang, P. Schaumont, and M. Orshansky, "New maskless debiasing method for lightweight physical unclonable function," in *Proceedings of the IEEE International Symposium on Hardware Oriented Security and Trust (HOST)*, McLean, VA, May 2017, pp. 134–139.

97. J. Katz, J.S. Shin, "Parallel and concurrent security of the HB and HB+ protocols," Cryptology ePrint Archive, Report 2005/461, 2005, http://eprint.iacr.org.

98. H. Gilbert, M. Robshaw, H. Silbert, "An active attack against HB+ – A provable secure lightweight authentication protocol," Cryptology ePrint Archive, Report 2005/237, 2005, http://eprint.iacr.org.

99. H.-Y. Chien, "SASI: A New Ultralightweight RFID Authentication Protocol Providing Strong Authentication and Strong Integrity", *IEEE Trans. Depend. Secure Comput.*, vol. 4, no. 4, pp. 337–340, 2007.

100. P. Peris-Lopez, J.C. Hernandez-Castro, J. Estevez-Tapiador, A. Ribagorda, "LMAP: a real lightweight mutual authentication protocol for low-cost RFID tags," *Printed Handout of Workshop on RFID Security –RFIDSec*, Graz, Austria, 06 July 2006.

101. P. Peris-Lopez, J.C. Hernandez-Castro, J. Estevez-Tapiador, A. Ribagorda, "M2AP: A minimalist mutual-authentication protocol for low-cost RFID tags," Lecture Notes in Computer Science, Springer, Berlin, 2006, pp. 912–923.

102. P. Peris-Lopez, J.C. Hernandez-Castro, J.M. Estevez-Tapiador, A. Ribagorda, "EMAP: An efficient mutual authentication protocol for low-cost RFID Tags," OTM Federated Conferences and Workshop: IS Workshop, IS'06, 4277 Lecture Notes in Computer Science, Springer, Berlin, 2006, pp. 352–361.

103. P. Peris-Lopez, J.C. Hernandez-Castro, J.M.E. Tapiador, A. Ribagorda "Advances in ultralightweight cryptography for low-cost RFID tags: Gossamer protocol," in *Proceedings of International Workshop on Information Security Applications*, Jeju Island, Korea 2008, pp. 56–68.

104. S. Vaudenay, "On privacy models for RFID," in *Proceedings of 13th International Conference on the Theory and Application of Cryptology and Information Security (ASIACRYPT '07)*, vol. 4833 of Lecture Notes in Computer Science, Springer, Kuching, Malaysia, December 2007, pp. 68–87.

105. S. Vaudenay, "Privacy models for RFID schemes," in *6th International Workshop, RFIDSec*, Istanbul, Turkey, June 2010, p. 65.

106. K. Ouafi and S. Vaudenay, Security and privacy in RFID systems [Ph.D. thesis], EPFL, Lausanne, Switzerland, 2011.

107. S. Canard, I. Coisel, J. Etrog, and M. Girault, "Privacy preserving RFID systems: Model and constructions," Cryptology ePrint Archive, Report 2010/405, 2010.

108. S. Canard, I. Coisel, and M. Girault, "Security of privacy preserving RFID systems," in *Proceedings of IEEE International Conference on RFID-Technology and Applications (RFID-TA 10)*, IEEE, Guangzhou, China, June 2010, pp. 269–274.

109. T. van Le, M. Burmester, and B. de Medeiros, "Universally composable and forward-secure RFID authentication and authenticated key exchange," in *Proceedings of the 2nd ACM Symposium on Information, Computer and Communications Security (ASIACCS '07)*, ACM, Singapore, March 2007, pp. 242–252.

110. M. Burmester, T. Van Le, B. De Medeiros, and G. Tsudik, "Universally Composable RFID Identification and Authentication Protocols," *ACM Trans. Inf. Syst. Secur.*, vol. 12, no. 4, Article 21, pp. 1–33, April 2009.

111. A. Juels and S.A. Weis, "Defining strong privacy for RFID," in *Proceedings of the 5th Annual IEEE International Conference on Pervasive Computing and Communications Workshops (PerCom '07)*, IEEE, New York, March 2007, pp. 342–347.

112. V. Chawla and D.S. Ha, "An Overview of Passive RFID," *IEEE Communications Magazine*, vol. 45, no. 9, pp. 11–17, September 2007.

Index

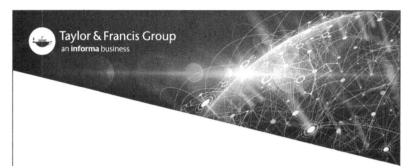

Taylor & Francis eBooks

www.taylorfrancis.com

A single destination for eBooks from Taylor & Francis
with increased functionality and an improved user
experience to meet the needs of our customers.

90,000+ eBooks of award-winning academic content in
Humanities, Social Science, Science, Technology, Engineering,
and Medical written by a global network of editors and authors.

TAYLOR & FRANCIS EBOOKS OFFERS:

A streamlined
experience for
our library
customers

A single point
of discovery
for all of our
eBook content

Improved
search and
discovery of
content at both
book and
chapter level

REQUEST A FREE TRIAL
support@taylorfrancis.com

Printed in the United States
by Baker & Taylor Publisher Services